Lenk, John D.

Complete guide to stereo
television (MTS/MCS)
troubleshooting repair

COMPLETE GUIDE TO STEREO TELEVISION (MTS/MCS) TROUBLESHOOTING AND REPAIR

COMPLETE GUIDE TO STEREO TELEVISION (MTS/MCS) TROUBLESHOOTING AND REPAIR

JOHN D. LENK

Consulting Technical Writer

Prentice Hall Englewood Cliffs, N.J. 07632

Library of Congress Cataloging-in-Publication Data

Lenk, John D.
  Complete guide to stereo television (MTS/MCS) troubleshooting and repair.

  Includes index.
  1. Television—Maintenance and repair. 2. Stereophonic sound systems—Maintenance and
repair. I. Title. TK6630.L47 1987   621.388 '87   87-12614
ISBN 0-13-160839-8

Editorial production supervision and
  interior design: *Ed Jones*
Manufacturing buyer: *Gordon Osbourne*

Printed in the United States of America
10 9 8 7 6 5 4 3 2 1

ISBN 0-13-160839 025

Prentice-Hall International (UK) Limited, *London*
Prentice-Hall of Australia Pty. Limited, *Sydney*
Prentice-Hall Canada Inc., *Toronto*
Prentice-Hall Hispanoamericana, S.A., *Mexico*
Prentice-Hall of India Private Limited, *New Delhi*
Prentice-Hall of Janpa, Inc., *Tokyo*
Prentice-Hall of Southeast Asia Pte. Ltd., *Singapore*
Editora Prentice-Hall do Brasil, Ltda., *Rio de Janeiro*

This book is dedicated to my wife Irene,
whose encouragement has made the book possible
and whose patience has made the work bearable.
And to our very special Lambie, be happy always.
You are a very good boy.
And to all who read this book,
Champagne Wishes
Caviar Dreams!

COMPLETE GUIDE TO COMPACT DISC (CD) PLAYER TROUBLE-SHOOTING AND REPAIR—1986

COMPLETE GUIDE TO LASER/VIDEODISC PLAYER TROUBLESHOOTING AND REPAIR—1985

COMPLETE GUIDE TO MODERN VCR TROUBLESHOOTING AND REPAIR—1985

COMPLETE GUIDE TO TELEPHONE EQUIPMENT TROUBLESHOOTING AND REPAIR—1987

COMPLETE GUIDE TO VIDEOCASSETTE RECORDER OPERATION AND SERVICING—1983

CONTENTS

PREFACE

The main purpose of this book is to provide a simplified, practical system of troubleshooting and repair for the many types of stereo television sets and stereo adapters now on the market. The book is the ideal companion to the author's many other troubleshooting/repair works listed after the title page of this book. However, there is no reference to the other works, nor is it necessary to have any other book to make full use of the information presented here. Of course, it is assumed that you are already familiar with the basics of television service. If not, you should not attempt troubleshooting or repair of any television set, stereo or monophonic!

It is virtually impossible to cover detailed troubleshooting and repair for all stereo TV sets and adapters in any one book. Similarly, it is impractical to attempt such coverage, since rapid technological advances soon make such a book's details obsolete. Very simply, you must have adequate service literature for any specific model of TV or adapter that you are servicing. You need schematic diagrams, part-location photos, descriptions of adjustment procedures, and so on, to do a proper troubleshooting/repair job.

Instead of trying to provide such details, this book concentrates on troubleshooting/repair approaches. This is done by providing a separate chapter for a cross section of stereo-TV circuits found in television receivers, hi-fi VCRs, and stereo-TV adapters.

In each chapter you will find (1) an introduciton that describes the purpose or function of the circuit (including specifications where applicable); (2) operating procedures for the stereo circuits (including a description of the controls, indicators, and interconnections in the case of stereo-TV adapters); (3) circuit descriptions or circuit theory (not generally available in the manufacturers service manuals); (4) sample test and adjustment procedures (including a full discussion of the test equipment required); and (5) a logical troubleshooting approach for the circuit (based on the manufacturer's recommendations).

Chapter 1 provides an introduction to stereo television, and includes a brief history, typical system configurations, encoding and transmission characteristics, reception and decoding characteristics, noise-reduction techniques, a glossary of terms, and a description of the author's troubleshooting approach.

Chapter 2 is devoted to the special test equipment needed for proper troubleshooting and repair of stereo-TV circuits. The chapter describes how controls and indicators on the test equipment can be used to isolate troubles in malfunctioning stereo-TV circuits. These tests are then referenced in later chapters.

Chapter 3 provides an overall description of Mitsubishi stereo-TV circuits, concentrating on the stereo decoders found in stereo-ready sets. This is followed by operating procedures, circuit descriptions, and test/adjustment procedures for the stereo circuits. The chapter concludes with step-by-step, circuit-by-circuit troubleshooting, based on trouble symptoms.

Chapter 4 provides coverage, similar to that of Chapter 3, for Mitsubishi hi-fi/stereo-VCR circuits.

Chapter 5 provides coverage, similar to that of Chapter 3, for Sony stereo-TV adapter circuits.

Chapter 6 provides coverage, similar to that of Chapter 3, for Sony stereo-TV circuits.

Chapter 7 provides coverage, similar to that of Chapter 3, for General Electric stereo-TV adapter circuits.

Throughout the various chapters, no attmept has been made to duplicate the full schematics for all circuits. Such schematics are found in the service literature for the particular equipment. Instead of a full schematic, the circuit descriptions are supplemented with partial schematics and block diagrams that show such important areas as signal flow paths, input/output, adjustment controls, test points, and power-source connections. These are the areas most important in troubleshooting.

By reducing full schematics to the essentials you will find the circuit easier to understand. You will also be able to relate circuit operation to the corresponding circuit of the stereo-TV equipment you are servicing.

Many professionals have contributed their talent and knowledge to the preparation of this book. The author gratefully acknowledges that the tremendous effort to make this book such a comprehensive work is impossible for one person, and wishes to thank all who have contributed, both directly and indirectly.

The author wishes to give special thanks to the following: Martin Plude' of B&K-Precision Dynascan Corportion; dbx Inc.; Publications Division, Hitachi Sales Corporation of America; Bob Stell and Carlos Nieves of General Electric; Everett Sheppard, Ron Smith, Walt Herrin, Robert Green, and Jeff Harris of Mitsubishi; Deborah Fee and Pat Wilson of N.A.P. Consumer Electronics (Magnavox, Sylvania, Philco); John Lostroscio of NEC Home Electronics; Thomas Lauterback of Quasar; Judith Fleming and J. W. Phipps of RCA; David Rhoades of Rhoades National Corporation; Andrea Messinger of Philip Stogel for Sansui; Donald Woolhouse of Sanyo; Theodore Zrebiec and Y. Shimazaki of Sony; Justin Camerlengo of Technics; and Matthew Mirapaul and John Taylor of Zenith.

The author extends his gratitude to Gloria Calagreco, Paul Becker, Matt Fox, Diane Spina, Melissa Halverstadt, Greg Burnell, Hank Kennedy, John Davis, Jerry Slawney, Irene Springer, Nancy Bauer, Wilma Curvino, Barbara Cassel, Karen Fortgang, Lisa Schulz, Pat Walsh, Ellen Denning, Linda Maxwell, Natalie Brenner, and Armond Fangschlyster of Prentice-Hall. Their faith in the author has given him encouragement, and their editorial/marketing expertise has made many of the author's books best-sellers. The credit must go to them. The author also wishes to thank Joseph A. Labok of Los Angeles Valley College for his help and encouragement throughout the years.

And to my wife, Irene Lenk, my research analyst, I wish to thank her. Without her help, this book could not have been written.

JOHN D. LENK

COMPLETE GUIDE
TO STEREO TELEVISION
(MTS/MCS)
TROUBLESHOOTING
AND REPAIR

1

INTRODUCTION
TO MTS (STEREO TV)

This chapter is devoted to the basics of the MTS system. Before we get into these basics, let us establish a few common terms.

MTS (multichannel television sound) is also called "stereo TV," "TV multichannel sound," "multichannel sound," "MCS," the "BTSC system," and even the "SAP system" in the literature. However, the terms "MTS" and "stereo TV" or "TV stereo" are now in the most common use (so we use these terms interchangeably throughout the book).

MTS or stereo TV is presently available in three basic versions. First, there are *stereo-ready* TV sets or receivers with built-in stereo capability (including dual loudspeakers and stereo amplifiers). Such TV sets have built-in *decoder circuits* to sense the presence of *stereo-TV broadcasts,* or *second audio program* (SAP) broadcasts and to produce the corresponding audio signals. Most major TV manufacturers include some stereo-TV/SAP sets in their line. Figure 1-1 shows some typical examples.

Second, there are *external stereo-TV decoders,* sometimes called *adapters.* Such decoders are essentially small "black boxes" (with or without controls) that make it possible to convert a monophonic (single audio channel) TV set into a stereo or SAP receiver. These external units contain the decoder circuits. However, it is generally necessary to provide an external system (speakers and amplifiers) to reproduce the stereo/SAP sound (although some decoders also provide amplification).

To use external decoders or adapters, there must be some provision on the TV set to tap the output of the TV tuner. [Typically, this is an MPX (multiplex) connector or jack on the rear panel of the TV set.] Figure 1-2 shows some examples of external decoders or adapters.

Third, there are TV sets described as "stereo adaptable" or as "ported for stereo." Such sets have built-in dual loudspeakers and stereo amplifiers but do not include the decoder. An external decoder must be used.

Typically, the stereo-adaptable or ported sets have an output connector from the tuner and audio input connectors to the stereo amplifier. The external decoder input is connected to the tuner output, while the decoder outputs are connected to the TV audio inputs. This eliminates the need for an external audio system (amplifier and speakers).

In addition to these three basic versions, most *high-fidelity* VCRs, and some stereo VCRs, include MTS or stereo-TV decoder circuits. This makes it possible for such VCRs to pick up, decode, and record MTS broadcasts and to play back such broadcasts in true stereo form.

Finally, in addition to the true stereo decoders, there are a number of devices that produce *simulated stereo* or that *enhance reception* of stereo TV. Figure 1-3 shows some examples of these devices, alternatively described as "stereo synthesizers," "enhancers," "processors," and "restorers."

In this book we are concerned primarily with troubleshooting and repair of true stereo-TV equipment, devoting a separate chapter to each device covered. Before we get into the specific circuit details, test/adjustment procedures, and troubleshooting, let us take a look at the MTS or stereo-TV picture.

(a)

(b)

(c)

(d)

(e)

(f)

FIGURE 1-1a. Sylvania RLD596PE Stereo Console (Courtesy of N.A.P. Consumer Electronics Corp.) (b). Magnavox RG4314R 20″ Stereo Color Table Model TV (Courtesy of N.A.P. Consumer Electronics Corp.) (c). Quasar Model PR552AW Projection Television (Courtesy of Quasar Company, Elk Grove Village, Illinois) (d). Hitachi CT2559 25″ Color Television (Courtesy of Hitachi Sales Corporation of America) (e). Mitsubishi CK-2561R Color Television (Courtesy of Mitsubishi Electric Sales America, Inc.) (f). General Electric 19CP6770 Stereo Television (Courtesy of General Electric Company).

FIGURE 1-2a Sony MLV-1100 Multichannel TV Sound Adapter (Courtesy of Sony Corporation of America)

FIGURE 1-2b General Electric SBA800 Multi-channel Sound Decoder (Stereo/ Bilingual Adapter) (Courtesy of General Electric Company)

FIGURE 1-3a Sansui Model AV-77 Audio-Video Processor. (Courtesy of Sansui Electronics Corporation.)

FIGURE 1-3b Rhoades TE-500 Stereo-Plexer. (Courtesy of Rhoades National Corporation.)

1–1. A BRIEF HISTORY OF MTS

In late 1978, the Broadcast Television System Committee (BTSC) of the Electronic Industries Association (EIA), on behalf of the television industry, formed a subcommittee for the purpose of formulating standards for the broadcasting and reception of multichannel television sound (MTS).

The proposed system was to include stereophonic as well as SAP program (second language, for example) enhancements of the main audio program. On December 22, 1983, the industry chose the *Zenith Transmission System,* coupled with the *dbx* Noise Reduction System,* and submitted the combined system (the BTSC system) to the Federal Communications Commission (FCC) on January 30, 1984.

The present BTSC system, introduced in 1984–1985, delivers a stereophonic program audio quality essentially limited only by the main channel (monophonic) quality, and a SAP audio program of slightly lesser quality.

1–2. TYPICAL MTS OR STEREO-TV SYSTEM

The hardware configuration for a typical television transmitter and a typical television receiver for the entire BTSC system is shown in Fig. 1–4. The multichannel signal baseband spectrum is shown in Fig. 1–5. Figure 1–6 shows the signal specifications in the BTSC standard transmission system.

1–2.1 Basic signal flow in MTS

The main channel signal is composed of the sum (L+R) signal of L (left) and R (right) signals and is the same as in the conventional TV sound specification. As a result, the same TV sound as in conventional broadcasting can be received by an ordinary TV set when MTS broadcasting is in effect. When the difference between the L and R signals (L−R) is transmitted, the L and R signals are played back from the main-channel signal (L+R) and the stereo signal (L−R) by the TV receiver.

The L−R signal is produced by amplitude-modulating a subcarrier with a frequency double the horizontal scanning frequency (fH) using the DSB-SC (doublesideband suppressed carrier) system. As a result, the frequency deviation of the main-carrier signal is double (50 kHz) that of the main-channel signal (L+R).

The carrier wave of the L−R signal is suppressed, so it is necessary to transmit a reference signal to demodulate the L−R signal correctly in the TV set. To do this, a signal with a frequency equivalent to the horizontal scanning frequency (fH), called the *pilot signal,* is inserted between the main-channel signal (L+R) and the stereo signal (L−R). The pilot signal is also used to indicate the presence or absence of the stereo signal (no pilot, no stereo).

---

*"dbx" is a registered trademark of dbx Inc.

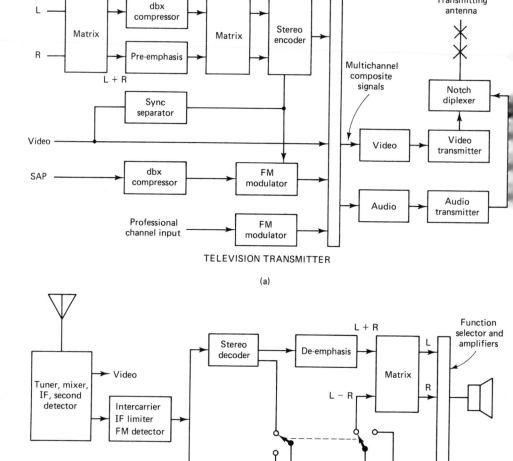

FIGURE 1-4   Hardware configuration for the BTSC system.

The SAP channel signal frequency-modulates a subcarrier with a frequency of 5fH and is locked to 5fH during nonmodulation.

The *nonpublic channel* (also called the *professional channel*) signal frequency-modulates the subcarrier with a frequency of 6.5fH. This channel is scheduled to be used for business, not for TV broadcasting.

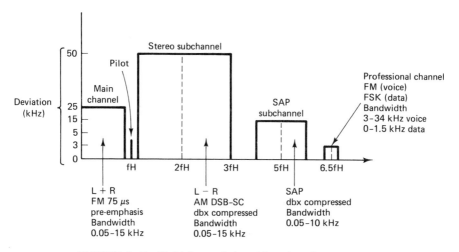

FIGURE 1-5 Multichannel signal baseband spectrum.

Service or Signal	Modulating Signal	Modulating Frequency Range (kHz)	Audio Processing or Preemphasis	Subcarrier Frequency[a]	Subcarrier Modulation Type	Subcarrier Deviation (kHz)	Audio Carrier Peak Deviation (kHz)
Monophonic	L + R	0.05-15	75 μs				25[b]
Pilot				fH	CW		5
Stereophonic	L - R	0.05-15	dbx compression	2fH	AM-DSBSC		50[b]
Second program	SAP	0.05-10	dbx compression	5fH	FM	10	15
Professional channel	Voice or data	0.3-34 voice; 0-1.5 data	150 μs	6.5fH	FM voice; FSK data	3	3

[a]fH = 15.734 kHz.
[b]Sum does not exceed 50 kHz.

FIGURE 1-6 BTSC standard signal specifications.

The sum signal (L + R), difference signal (L − R), pilot signal, SAP channel signal, and the nonpublic channel signal are added to produce the *multichannel composite signal,* which, in turn, FM-modulates the transmitted TV broadcast signal. A *noise-reduction system,* which specifically encodes and transmits the L − R and SAP signals, is used to reduce noise in the TV set.

1-2.2 MTS signal characteristics

As shown in Fig. 1-5, the main-channel modulation consists of an (L + R) audio signal. The preemphasis is 75 μs. The L − R audio is subjected to level encoding, which is part of the dbx companding system (Sec. 1-5) that includes complementary decoding (expansion) in the reciever. The encoded L − R signal causes double-sideband suppressed-carrier amplitude modulation of a subcarrier and 2fH. The audio bandwidth of preemphasized L + R, and of encoded L − R, is 15 kHz.

The main-channel peak deviation is 25 kHz. With level encoding temporarily replaced by 75-μs preemphasis, the subchannel peak deviation is 50 kHz. When L and R are statistically independent, the peak deviation of the main channel and the stereophonic subchannel combined is also 50 kHz (due to the interleaving property). When L and R signals are not statistically independent, or when (L + R) and (L − R) signals do not have matching preemphasis characteristics (for example, when L − R is encoded), the combined deviation of the main channel and the stereophonic subchannel is limited to 50 kHz, and the separate components assume their respective natural levels. A CW (continuous wave) pilot subcarrier signal of frequency fH is transmitted with a main-carrier deviation of 5 kHz.

The subcarrier for the SAP channel has a frequency of 5 fH (78.670 kHz) and is frequency-locked to 5fH in the absence of modulation. The SAP audio signal is subjected to level encoding identical to that of the L − R signal. The resulting SAP modulating signal is bandlimited to 10 kHz, and frequency-modulates the SAP subcarrier to a peak deviation of 10 kHz. The main carrier deviation by this subcarrier is 15 kHz.

The *professional subchannel* has a subcarrier located at approximately 6.5 fH and modulates the main carrier by a peak deviation of 3 kHz.

The foregoing descriptions illustrate the case for a fully loaded MTS baseband, as shown in Fig. 1-5. Some transmissions might consist of monophonic audio and SAP, with or without nonpublic subcarriers, as shown in Fig. 1-7.

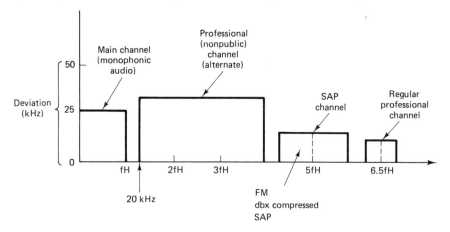

FIGURE 1-7   Baseband with monophonic audio and SAP, with or without professional (nonpublic) subcarriers.

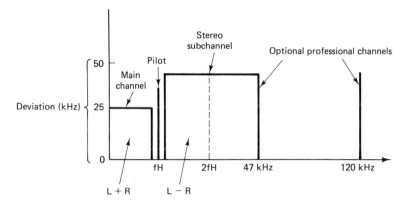

FIGURE 1-8 Baseband with L+R, pilot, L−R, and optional professional (nonpublic) subcarriers.

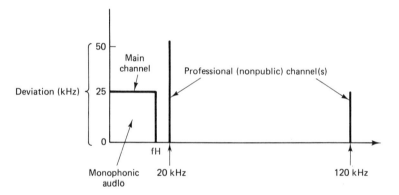

FIGURE 1-9 Baseband with monophonic audio and optional professional (nonpublic) subchannels.

Another possible baseband configuration might consist of main (L+R) channel, pilot, and stereophonic (L−R) subchannels, with or without nonpublic subcarriers, as shown in Fig. 1-8.

The last baseband configuration, consisting of monophonic audio, with or without nonpublic subcarriers, is shown in Fig. 1-9. This configuration is essentially the same as that of conventional TV broadcasts (without MTS).

1-2.3 Audio signal processing and companding

The dbx companding system (Sec. 1-5) is the mandatory noise- and interference-reduction companion to the Zenith transmission system, and is based on complementary audio processing at the transmitter and receiver. (Noise reduction is essential to get suitable stereo signal-to-thermal-noise ratios.)

To maintain monophonic (L+R) compatibility, the main channel is not companded. Only the subchannels are companded. However, this still produces satisfactory noise reduction since most of the noise is introduced in the L−R and SAP subchannels.

The dbx system includes a *stereo-difference compressor,* and a separate *SAP compressor,* at the transmitter. A *single expander* for both stereo and SAP is provided in the receiver. Stereo separation is influenced by the degree to which compressor (transmitter) and expander (receiver) processing are complementary.

1–3.   ENCODING AND TRANSMISSION OF MTS

Although we are concerned with reception in this book, it is helpful to understand the encoding and transmission process used by stereo TV broadcast stations. For that reason, we describe (in brief) the basic stereo generator, basic SAP generator, composite signal, and transmitter characteristics for stereo TV or MTS.

1–3.1 Basic stereo generator

The block diagram of the basic stereo generator is shown in Fig. 1–10. The stereo generator creates the stereo portion of the composite signal from L and R audio inputs. Input low-pass filters (LPFs) on the L and R inputs remove components above 15 kHz to prevent overload of the circuits. The L+R and L−R signals are made by a matrix circuit. Ths phase response and amplitude response of the L+R and L−R paths are matched to preserve separation.

The L+R is preemphasized by 75 μs, and the L−R is compressed by a dbx compressor. Clippers, followed by closely matched sharp-cutoff 15-kHz LPFs are included in both the L+R and L−R paths to prevent overmodulation and crosstalk into other portions of the baseband composite signal.

The filter in the L−R path is essential for removing potentially large out-of-band components generated in the compressor. If these components are allowed

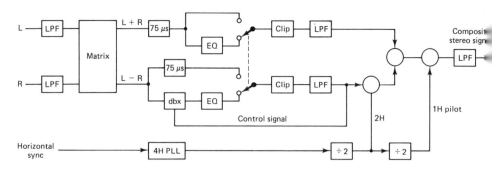

FIGURE 1–10   Basic stereo generator.

to pass, the monophonic and SAP signal-to-noise (S/N) ratios would be degraded. Also included in the L + R and L − R paths are equalizers (EQs) which compensate for any phase and amplitude errors in the compressor.

Switching is provided to substitute the compressor and equalizers with a matched 75-μs preemphasis in the L − R path for testing the transmission system. Matched preemphasis networks are needed so as not to deteriorate separation during testing.

To ensure good tracking between the compressor and expanders in the TV receivers, the control line for the compressor is taken from the output of the L − R LPF. The control signal in the compressor and expanders are taken from as close to the same signal as possible, for the best possible tracking.

The compressed L − R signal is modulated (AM-DSBSC) and summed to the preemphasized L + R signal. The modulating subcarrier is at 2H and derived from a PLL (phase-locked loop) locked to horizontal sync. The PLL also supplies the pilot which is added to the signal. The resultant signal is filtered by an LPF (to protect other subcarrier services from spillover) before appearing as the composite stereo signal. To preserve separation, this filter must be phase linear and flat to 46.5 kHz.

1–3.2 Basic SAP generator

Audio processing for the SAP channel is shown in Fig. 1–11. This processing is the same as for the L − R signal in the stereo generator, with two exceptions. Phase compensation is not needed, and the LPFs have 10-kHz cutoffs instead of 15 kHz.

A 5H subcarrier is frequency-modulated by the processed SAP audio. Peak deviation of the subcarrier is 10 kHz. The modulated subcarrier then passes through a bandpass filter designed to protect adjacent subcarrier services from spillover (by preserving the phase and amplitude of the FM sidebands to minimize distortion).

The 5H subcarrier is locked to the fifth harmonic of H during moments of audio silence. Figure 1–11 shows one implementation where this is done with a PLL circuit. The processed audio is added to the control voltage of a VCO (voltage-

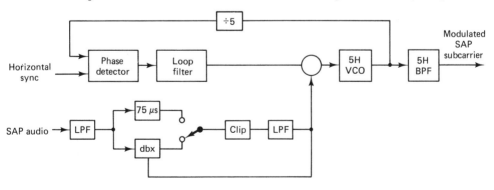

FIGURE 1–11   Basic SAP generator.

controlled oscillator) that is part of a 5H PLL locked to H. In the configuration of Fig. 1–11, frequency lock and phase lock are achieved only during audio silence. Frequency lock is maintained only for high modulating frequencies, above 1.2 kHz, and the loop unlocks for high levels of low-frequency audio.

1-3.3 Composite signal

The output of the SAP generator is added to the output of the stereo generator to form a composite signal. This composite signal is used to frequency-modulate the audio portion of the TV broadcast signal.

1-3.4 Basic transmitter considerations

TV transmitters used for broadcast of MTS or stereo TV generally have two special functions in addition to the basic broadcast functions. Such transmitters must (1) minimize the ICPM (incidental carrier-phase modulation) of the video transmission, and (2) provide a *notched diplexer* to prevent spillover of the video sideband components into the audio carrier spectrum. These functions help to reduce *audio buzz* from the transmitter.

1-4. RECEPTION AND DECODING OF MTS

In the remaining chapters, we describe operation of the various circuits used by TV set manufacturers to receive and decode stereo/SAP signals. To give you the necessary background for understanding these circuits, let us look at the basic characteristics of the MTS receiver, 4.5-MHz FM detector, stereo decoder, expander, and SAP decoder used in stereo/SAP TV sets.

1-4.1 MTS receiver characteristics

High-quality stereo sound from TV requires improvements in thermal-noise performance and in intercarrier buzz over those found in conventional TV receivers. The improvement in noise performance is obtained through dbx companding, which also contributes to the reduction of intercarrier buzz (and distortion).

Intercarrier buzz can be reduced (but not totally eliminated) by lowering transmitter ICPM (as discussed in Sec. 1–3.4). In the receiver, such buzz can be minimized through an *intercarrier detection process* found in *quasi-split-sound* or *quasi-parallel-sound* receivers, shown in Fig. 1–12.

Two other types of TV receivers, *separate-sound* and *split-sound* receivers, are unsatisfactory for MTS. Separate-sound receivers require an extra tuner, and split-sound receivers can produce buzz. (Such buzz can be caused in the tuner, or before the tuner, as may happen in a cable system, and is not eliminated in a true split-sound audio second detector.)

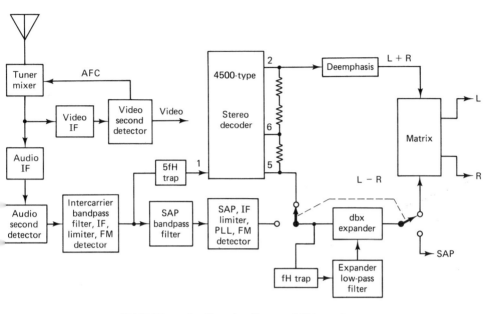

FIGURE 1-12   Quasi-split-sound TV receiver.

1-4.2   MTS 4.5-MHz FM detector characteristics

The 4.5-MHz intercarrier FM detector found in MTS sets requires improved performance over that of the FM detector in monophonic sets. The primary reason for this is that the MTS baseband signal occupies approximately 90 kHz, including several subcarriers. As a result, 180 kHz of bandwidth is needed in the input bandpass filter of the MTS 4.5-MHz intercarrier circuit. In most MTS receivers, a doubly tuned or equivalent ceramic filter provides adequate bandwidth and selectivity.

The FM detector of an MTS receiver must be widened and linearized (compared to a conventional TV) to prevent intermodulation, which can result in crosstalk between the SAP and stereo signal. An improved signal-to-noise (S/N) ratio is also required for MTS. Such an improvement in S/N is provided by the dbx companding. However, because the MTS receiver must also be compatible with monophonic broadcasts (which do not normally have dbx companding), the $L+R$ signal is not companded. As a result, the L and R audio noise and buzz levels are determined by the S/N ratio of the $L+R$ channel signal.

1-4.3   Stereo decoder characteristics

Figure 1-12 shows a 4500-type stereo decoder used in some MTS TV sets. The L and R outputs are not deemphasized but are re-matrixed to supply the required $L+R$ and $L-R$ outputs. No matter what type of stereo decoder is used, the decoder must have several features.

One feature is immunity of the pilot detector circuit against signals in the SAP carrier. This prevents disturbing the pilot phase and eliminates the need for an external 5fH trap. Another desirable feature for a stereo decoder is a narrow acquisition range for the pilot PLL. A third feature is individual (L + R) and (L − R) outputs. A fourth highly-desirable feature is the elimination of the pilot from the L − R output. This prevents the pilot from contributing to the expander control signal (which would cause mistracking and could result in separation loss). Also, elimination (or at least considerable attentuation) of the pilot from the L + R output prevents the pilot from being heard. (Unless the pilot is definitely eliminated from L + R, it is possible that the pilot frequency of 15.734 kHz can be reproduced by some stereo-system loudspeakers, particularly the tweeters.)

1–4.4   Expansion characteristics

A number of demodulation products are present at frequencies above the compressed L − R signal. Included are modulated-related AM sidebands around 2fH and 4fH, together with SAP products located at 3 fH, 5 fH, and 7 fH. Some remainder of the pilot may also be present. The expander control circuit is sensitive to all of these demodulation products. This is the reason for the expander low-pass filter shown in Fig. 1–12.

Good stereo separation requires both amplitude matching and phase matching between the deemphasized L + R and the expanded L − R signals. (As a rule of thumb, an overall response difference of 0.3 dB in amplitude and 3 ° in phase produces 30 dB of stereo separation.) The expander circuits of some stereo-decoder ICs can introduce some phase distortion (which needs equalization if good stereo separation is to be retained). This is the reason for phase-equalization networks in both the L − R and L + R channels of some MTS receivers.

1–4.5   SAP decoder characteristics

As shown in Fig. 1–12, the SAP FM subcarrier is separated from the composite baseband signal in the SAP bandpass filter. The SAP channel at 5fH is so far removed from the L − R channel that very little filtering is needed. However, a disturbance known as "buzz beat" can result when intercarrier components beat with the instantaneous FM carrier frequency. A typical solution is to include narrow traps at 4fH and 6fH in the SAP bandpass filter.

The SAP IF, limiter, and FM discriminator circuits are similar to corresponding circuits in conventional TV receivers. For example, the SAP carrier can be detected by a pulse-counting detector or by a PLL used as a discriminator. (The PLL discriminator is most popular.)

A *carrier threshold* circuit is required in the SAP channel. When there is no SAP broadcast, the threshold circuit can be used to mute the channel, to switch to mono or stereo, or to protect the SAP channel from signals other than SAP transmission.

In virtually all MTS receivers, the dbx expander is switched between the $L-R$ signal path and the SAP path. Although stereo and SAP can be received simultaneously, it is assumed that only one is used at a time.

1-5. THE dbx NOISE-REDUCTION SYSTEM FOR MTS

dbx noise reduction is an essential part of the MTS or stereo TV system. In the simplest of terms, portions of the transmitted signal are compressed (or encoded) at the TV transmitting station. These same signals are expanded (or decoded) at the TV set. The process is known as companding (or compandoring, in some literature).

The following is a brief discussion of the companding process, covering both the "why" and the "how." Figures 1-13 and 1-14 show simplified block diagrams of the compressor and expander circuits, respectively.

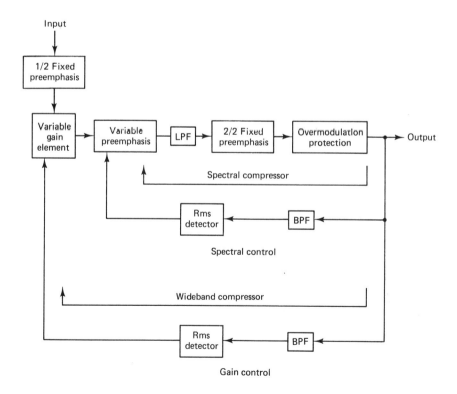

FIGURE 1-13 Basic compressor circuit.

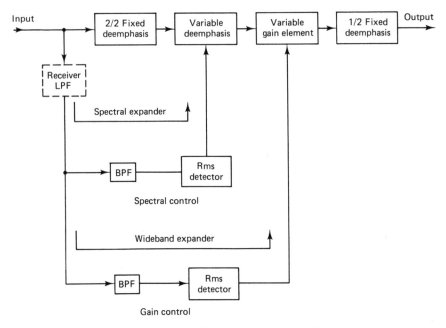

FIGURE 1-14   Basic expander circuit.

1-5.1   The MTS transmission system

As discussed, the MTS system transmits the sum of the left and right stereo audio signals (L + R) in the spectrum space usually occupied by the conventional or mono TV audio signal (Fig. 1-5). The stereo information is encoded by subtracting the right audio signal from the left (L − R) and transmitting the difference via an AM subcarrier located at 31.468 kHz (twice the video horizontal-scanning frequency of 15.734 kHz), superimposed on the conventional FM audio carrier.

1-5.2   Noise reduction

Because the L − R or stereo subcarrier is at a higher frequency than the L + R signal, stereo reception is about 15 dB noisier than mono reception, even under ideal conditions. Typically, the stereo *S/N* ratio is about 50 dB. The SAP subcarrier (at an even higher frequency of 78.67 kHz) has a typical *S/N* ratio of 33 dB. The practical effect of this noise is to reduce the coverage area for both stereo and SAP, compared to the mono coverage area (under identical conditions).

The dbx noise-reduction system is designed to improve the *S/N* ratios. (Theoretically, the goal is to eliminate any noise increase when going from mono to stereo, and to make SAP "listenable.")

Note that noise reduction is added only to the L − R and SAP channels, leaving the mono L + R signal unchanged. This ensures compatibility with conventional TV broadcasts. (If you have a conventional TV set, you will not know that MTS is being broadcast.)

Identical companding is used in both the L – R and SAP channels. This allows a single noise-reduction circuit to be switched between the L – R and SAP channels in the TV receiver.

Obviously, the dbx noise-reduction system must be especially powerful. The system must encode (compress) the audio signal in such a way that the signal itself consistently masks the noise of the channel during transmission, and must then decode (expand) the transmitted signal to recover the original audio. This requires that the transmitted signal be at a consistently high amplitude *and* that the spectrum contain substantial high-frequency content. This miracle is done by preemphasis, deemphasis, spectral companding, and wideband-amplitude companding, all of which we discuss next.

1–5.3 Preemphasis and deemphasis

The dbx noise-reduction system uses both preemphasis and deemphasis. The preemphasis consists of a zero at 72.7 μs, a zero at 390 μs, and a pole at 30 μs. This provides a steep section of the frequency-response curve between 2 and 5.5 kHz to help the dbx noise-reduction system to overcome the large amounts of noise present. In the TV receiver, a corresponding deemphasis restores correct tonal balance to the program material and reduces "hiss" picked up in transmission.

1–5.4 Spectral companding

The dbx noise-reduction system includes a stage where preemphasis is varied to suit the signal. This function is called *spectral companding*. When very little high-frequency information is present in the audio, the spectral compressor (Fig. 1–13) provides large high-frequency preemphasis. When strong high frequencies are present, the spectral compressor provides deemphasis, thus reducing the potential for high-frequency overload. The transmitted signal is dynamically adjusted to have high-frequency content to provide good masking.

In the receiver (Fig. 1–14) the spectral expander restores high-frequency signals to their proper amplitude. The expander also attenuates high-frequency noise when little high-frequency information is present. When strong high-frequency signals are present, the signal itself masks the noise.

1–5.5 Wideband-amplitude companding

Wideband-amplitude companding is the third stage or function of dbx noise reduction. Such companding is responsible for keeping the signal level in the transmission channel high at all times.

The compressor (Fig. 1–13) reduces the dynamic range of input signals by a factor of 2 : 1 in decibels. The transmitted signal level tends toward about

14% modulation, which allows transient peaks to overshoot without causing overmodulation.

During reception (Fig. 1-14), the wideband-amplitude expander restores signal levels to their proper amplitude. When the signal is low in amplitude, the decoder attenuates channel noise toward the point of inaudibility. During high-amplitude passages, the signal masks the noise.

1-6. GLOSSARY OF TERMS

Following are some terms commonly used in MTS or stereo TV literature. Although these terms are used primarily in engineering, the terms are also of value to the service technician.

Companding (or compandoring): A noise-reduction process used in the stereo subchannel and the SAP subchannel, consisting of encoding (compression) before transmission, and decoding (expansion) after reception.

Composite stereophonic baseband signal: The stereophonic sum modulating signal, the stereophonic difference encoded signal, and the pilot subcarrier.

Crosstalk: An undesired signal occurring in one channel caused by an electrical signal in other channels.

Decibel ERMS value: The exponentially time-weighted root-mean-square (erms) value converted to decibels as follows:

$$\text{decibel ERMS value} = 20 \log 10 \frac{\text{ERMS value}}{\text{reference}}$$

where "reference" is the 0-dB value.

Encoding: See Companding.

Equivalent input separation: A method of specifying the stereophonic separation by referring variations from ideal at the output, back to the input. To do this, an input signal that causes a nonideal output is varied by degrading input separation until the output conforms to the ideal. The amount of input separation degradation required is the equivalent input separation.

Equivalent input tracking: A method of specifying the tracking ability of the encoding process by referring variations from ideal at the output, back to the input of the encoder. To do this, an input signal that causes a nonideal output is varied until the output conforms to the ideal. The amount of input variation required is the equivalent input tracking.

Equivalent modulation: See Equivalent input tracking.

Incidental carrier-phase modulation (ICPM): Angle modulation of the video carrier by video-signal components which, when detected in TV receiver intercarrier circuits, cause an audio interference known as "intercarrier buzz."

Left (or right) audio signal: The electrical output of a microphone or combination of microphones placed so as to convey the intensity, time, and location of sounds originating predominantly to the listerner's left (L) (or right R) of the center of the performing area.

Left (or right) stereophonic channel: The transmission path for the left (or right) audio signal.

Main channel: The band of frequencies from 50 to 15,000 Hz which frequency-modulates the audio carrier.

Multichannel sound: Multiplex transmission on the TV audio carrier.

Multiplex transmission: The simultaneous transmission of the TV program main channel audio signal and one or more subchannel signals. The subchannels include a stereophonic subchannel, a second audio program or SAP subchannel, a non-program-related subchannel, and a pilot subcarrier. (The FCC does not require or restrict the use of the SAP subchannel or the non-program-related subchannel.)

Non-program-related subchannel: The subchannel for the multiplex transmission of a frequency-modulated subcarrier for telemetry or other purposes.

Pilot subcarrier: A subcarrier serving as the control signal for use in the reception of TV stereophonic sound (MTS) broadcasts.

Second audio program (SAP) broadcast: The multiplex transmission of a second audio program using the second audio program subchannel.

Second audio program (SAP) subchannel: The channel containing the frequency-modulated second audio program subcarrier.

Second program audio signal (also called the second language audio signal in some literature): The monophonic audio signal delivered to the SAP encoder.

Second program encoded signal: The second program audio signal after encoding.

Seventy-five-microsecond equivalent modulation: The audio signal level prior to encoding that results in a stated percentage modulation when the encoding process is replaced by 75-μs preemphasis.

Spectral compression: A process where variations in spectral constant of an audio signal are reduced by varying a frequency-filtering function applied to the signal in response to variations in spectral content of the signal.

Stereophonic difference audio signal: The left audio signal minus the right audio signal (L − R).

Stereophonic difference encoded signal: The stereophonic difference audio signal after encoding.

Stereophonic separation: The ratio of the electrical signal caused in the right (or left) stereophonic channel to the electrical signal caused in the left (or right) stereophonic channel by the transmission of only a right (or left) channel.

Stereophonic sum audio signal: The left audio signal plus the right audio signal (L + R).

Stereophonic sum channel compensation: A process where the phase and amplitude response resulting from bandlimiting in the process of encoding the stereophonic different audio signal (which, uncompensated, would detrimentally affect stereo separation) is compensated by an identical phase and amplitude response applied to the stereophonic sum audio signal.

Stereo sum modulating signal: The stereophonic sum audio signal after compensation, preemphasis, and other processing.

Stereophonic subcarrier: A subcarrier having a frequency which is the second harmonic of the pilot subcarrier frequency and which is used in TV stereophonic sound broadcasting (MTS).

Stereophonic subchannel: The subchannel containing the stereophonic subcarrier and associated sidebands.

Wideband-amplitude compression: A process where the synamic range of an audio signal is compressed by simultaneously varying the gain of all audio frequencies equally.

1-7. THE TROUBLESHOOTING APPROACH

Now that we know all about MTS, stereo-TV, and SAP, let us get into some actual circuits, see how they work, how to test and adjust them, and how to pinpoint troubles (hopefully in the most efficient way). Before we get to the details, here are some general points to consider.

1-7.1 Basic TV/VCR troubleshooting and repair

It is assumed that you are already familiar with TV/VCR service procedures (both color and black/white), including the use of test equipment, safety precautions (leakage current checks, handling electrostatically sensitive or ES devices, high-voltage checks at the picture tube, etc.), installation, routine maintenance, operating procedures, and basic troubleshooting (including solid-state troubleshooting). If not, and you plan to service MTS or stereo-TV sets, you are in terrible trouble. You had better read the author's best selling *Handbook of Simplified Television Service* (Englewood Cliffs, N.J.: Prentice-Hall, Inc., 1977), *Handbook of Advanced Troubleshooting* (Englewood Cliffs, N.J.: Prentice-Hall, Inc., 1983), *Complete Guide to Videocassette Recorder Operation and Service* (Englewood Cliffs, N.J.: Prentice-Hall, Inc., 1983), and *Complete Guide to Modern VCR Troubleshooting and Repair* (Englewood Cliffs, N.J.: Prentice-Hall, Inc., 1985).

1-7.2 Special test equipment

Although most of the test and adjustment procedures for MTS or stereo-TV can be performed using conventional TV/VCR/audio test equipment (meters, oscilloscopes, counters, signal generators, probes, etc.) you need a *special-purpose signal generator* to service the stereo decoder circuits properly. This generator must be capable of producing or simulating the stereo-TV and SAP signals transmitted by an MTS television station. Because of the importance of such generators, we devote all of Chapter 2 to the subject of special test equipment and procedures for MTS.

1–7.3 Circuit coverage

Most of this book is devoted to descriptions of the stereo decoder circuits found in stereo-TV sets, hi-fi, VCRs, and stereo adapters. If you understand operation of these circuits, you should have no difficulty in servicing the decoder portion of MTS TV equipment. (Circuit descriptions just happen to be that portion most often omitted in the service manuals for MTS equipment. They simply assume that you know how the circuits work, just as you do television and audio circuits.)

The operating/adjustment/test procedures found in MTS service literature are generally very good. As a result, we do not dwell on such procedures. Also, we do not cover all circuits in the TV set or VCR, but concentrate on those circuits that make stereo TV different from conventional or mono TV. Generally, for a TV set, this means starting from the IF output of the tuner and tracing through to the dual speakers. For a hi-fi VCR, this means tracing through the stereo-decoder circuits and noting how the circuit interact with the remaining VCR circuits. If you digest this information for a number of stereo-ready TV sets, stereo decoder/ adapters, and hi-fi VCRs, you should instantly recognize corresponding circuits in any MTS equipment.

1–7.4 Chapter format (how to use the book)

Throughout the remainder of the book (Chapters 3 through 7), a separate chapter is devoted to each equipment model. A generally consistent format is used in each chapter for your convenience. Each chapter starts with overall descriptions (including specifications where applicable), followed by operating procedures, circuit descriptions, and test/adjustment procedures. Each chapter concludes with troubleshooting tips and approaches for that specific equipment. Operating/adjustment/test procedures are omitted where repetitive.

1–7.5 Operating/adjustment/test procedures

The procedures found in each chapter apply only to the circuits described in the same chapter. Keep in mind that the procedures are the only procedures recommended by the manufacturer for that particular model of MTS equipment. Other manufacturers may recommend more or less adjustment and test. It is your job to use the correct procedures for the equipment you are servicing.

Also remember that some disassembly and reassembly are probably required to reach test/adjustment points. Although we do show the electrical locations for the adjustment/test controls and measurement points (test points, or TPs), we do not include any disassembly/reassembly, for two reasons.

First, disassembly/reassembly proceudres are unique and can apply to only one model of equipment. More important, disassembly and reassembly (both electrical and mechanical) is one area where the MTS service literature is generally well written and illustrated. Just make sure that you observe all the notes, cautions, and warnings found in the disassembly/reassembly sections of the service literature.

1-7.6 Circuit troubleshooting

After studying the circuits and circuit descriptions found in each chapter, you should have no difficulty in understanding the schematic and block diagrams of similar MTS equipment. This understanding is essential for logical troubleshooting and service, no matter what type of electronic equipment is involved. No attempt has been made to duplicate the full schematics for all circuits. Such schematics are available in the service literature for the particular equipment.

Instead of a full schematic, the circuit descriptions are supplemented with partial schematics and block diagrams that show such important areas as signal flow paths, input/output, adjustment controls, test points, and power-source connections. These are the areas most important to service and troubleshooting. By reducing the schematics to these areas, you will find the circuit easier to understand, and you will be able to relate circuit operation to the corresponding circuit of the MTS equipment that you are servicing.

1-8. THE BASIC TROUBLESHOOTING FUNCTIONS

On the off-chance that you do not know the author's basic troubleshooting approach (for any electronic equipment), here is a summary. Troubleshooting can be considered as a step-by-step logical approach to locate and correct any fault in the operation of equipment. In the case of MTS equipment (stereo-TV sets, hi-fi VCRs, and stereo decoder/adapters), seven basic functions are required.

First, you must study the equipment using service literature, user instructions, schematic diagrams, and so on, to find out how each circuit works when operating normally. In this way, you will know in detail how given equipment should work. This is why the theory of operation for a cross section of MTS equipment is included in Chapters 3 through 7.

The functions and features of all MTS equipment are similar, but not identical, to those of all other MTS equipment. If you do not take the time to learn what is normal, you will never be able to distinguish what is abnormal.

For example, in some stereo-TV sets, the decoder functions (sometimes referred to as the *multiplex,* or MPX, functions) are contained in a single IC, with a few additional circuits. In other configurations, several ICs (typcially, three to five) are used in the decoder circuits.

Similarly, it is common practice to place all of the decoder functions on a single PC (printed circuit) board (generally known as the decoder board or multiplex board). In other configurations, some of the usual audio functions are included on the same board. Only a careful study of the schematics and block diagrams in the service literature will show such details.

Second, you must know the function of, and how to manipulate, all controls associated with MTS. This is why the operating controls and indicators for the MTS functions are included in Chapters 3 through 7. Again, you must learn the operating controls for the equipment being serviced.

For example, in some stereo-TV sets, you must operate two controls or switches to get SAP (one switch to select stereo/SAP instead of mono, and another switch to select either stereo or SAP). In other sets, the stereo/SAP decoder circuits switch from mono to stereo automatically whenever a stereo or SAP signal is tuned in. Then you select stereo or SAP as desired. These are functions you must understand (and patiently explain to customers) for each set you service.

Also, the number of stereo/SAP front-panel indicators can be quite different for various stereo-TV sets. In some cases, there is only one indicator (labeled "stereo" or "MTS") that turns on automatically when either stereo or SAP (or both) are tuned in.

In other configurations, there are two indicators (one for stereo and another for SAP) that turn on when the corresponding broadcast is tuned in. Going further, some stereo-TV sets have three indicators. Two of the indicators (stereo and SAP) turn on to indicate the presence of corresponding broadcast signals, while the third indicator turns on to indicate which function has been selected.

As you can see, it is difficult, if not impossible, to check out MTS equipment without knowing how to set the controls. Besides, it makes a bad impression on the customer if you cannot find the stereo/SAP switch, especially on the second service call.

Third, you must know how to interpret service literature and how to use test equipment. Together with good test equipment that you know how to use, well-written service literature is your best friend. In general, MTS service literature is good as far as procedures and drawings are concerned. Unfortunately, MTS literature is often weak when it comes to descriptions of how circuits operate (theory of operation). The "how it works" portion of most player literature is often sketchy, or simply omitted, on the assumption that you and everyone else know MTS theory as well as circuit functions.

Fourth, you must be able to apply a systematic, logical procedure to locate troubles. Of course, a "logical procedure" for one type of equipment is quite illogical for another.

Fifth, you must be able to analyze logically the information of an *improperly operating MTS equipment.* For that reason, much of the troubleshooting information in this book is based on trouble symptoms and their relation to a particular circuit or group of circuits.

The information to be analyzed may be in the form of performance (such as failure of the stereo or SAP indicator to turn on when a known-good broadcast signal is available). The information can also be analyzed in the form of response to test equipment signals (such as waveforms or signals monitored with an oscilloscope when inputs signals from the special-purpose generator described in Chapter 2 are applied to the MTS circuits).

Either way, it is *your analysis* of the information that makes for logical, efficient troubleshooting.

Sixth, you must be able to perform complete checkout procedures on MTS equipment that has been repaired. The checkout may be only simple operation, such as selecting either stereo or SAP (or both) in turn, and checking response.

While on the subject of checkout, keep the following points in mind. MTS equipment is essentially audio or stereo equipment. When an MTS device is used in conjunction with external stereo equipment, it is assumed that you know how to operate the controls of the stereo system to amplify and reproduce the MTS circuit output. An improperly adjusted stereo system can make a perfectly good MTS circuit appear to be bad. For example, if the graphic equalization controls are set to some weird combination, any MTS circuit can appear equally weird. One suggestion for evaluation of an MTS circuit *in the shop* is to have at least one stereo system of known quality. All MTS equipment (hi-fi VCRs and TV sets with external stereo amplification) passing through the shop can be compared against the same standard.

Another point to consider is that some MTS equipment simply produces better sound than other equipment (just as is the case with stereo equipment). For example, frequency response, dynamic range, and signal-to-noise ratio are greater for one type of equipment. You can waste hours of precious time (money) trying to make the inferior equipment perform like the quality components, if you do not know what is "normal" operation. This is especially important when working in audio equipment, where all customers claim to have a "golden ear."

A complete checkout of MTS equipment can also involve adjustment of the decoder circuits (and possibly the related audio circuits). This brings up a problem. Although adjustment or setting of controls (both internal and front-panel) can affect circuit operation, such adjustment can also lead to false conclusions during troubleshooting. There are two extremes taken by some technicians during adjustment.

On one hand, the technician may launch into a complete alignment procedure once the trouble is isolated to a circuit. No control, no matter how inaccessible, is left untouched. The technician reasons that it is easier to make adjustments than to replace parts. While such a procedure eliminates improper adjustment as a possible fault, the procedure can also create more problems than are repaired. Indiscriminate adjustment is the technician's version of "operator trouble."

At the other extreme, a technician may replace part after part where a simple screwdriver adjustment will repair the problem. This usually means that the technician simply does not know how to perform the adjustment procedure or does not know what the control does in the circuit. That is why we start each adjustment procedure with a description of what is being done by the control.

To take the middle ground, do not make any internal adjustments during the troubleshooting procedure until trouble has been isolated to a circuit, and then only when the trouble symptom or test results indicate possible maladjustment. This middle-ground approach is taken throughout this book.

In any event, some checkout is required after any troubleshooting. One reason is that there may be more than one problem. For example, an aging part may cause high current to flow through a resistor, resulting in burnout of the resistor. Logical troubleshooting may lead you quickly to the burned-out resistor, and replacement of the resistor restores operation. However, only a thorough checkout can reveal the original high-current condition that caused the burnout. Another reason for after-service checkout is that the repair may have produced a condition that requires readjustment (such as after replacement of a major IC in the decoder circuits).

Seventh, you must be able to use the proper tools and test equipment. Fortunately, the tools and test equipment used in MTS circuits are the same as for TV and VCR service. The only exception is the special-purpose generator described in Chapter 2.

In summary, before starting any troubleshooting job, ask yourself these questions:

Have I studied all available service literature to find out how the specific equipment works, including any related circuits?

Can I operate the equipment properly, including any special control settings?

Do I really understand the service literature, and can I use all required test equipment and tools properly?

Using the service literature and/or previous experience on similar equipment, can I plan out a logical troubleshooting procedure?

Can I analyze logically the results of operating checks, as well as checkout procedures involving test equipment?

Using the service literature and/or experience, can I perform complete checkout procedures on the equipment, including adjustments if necessary?

Once I have found the trouble, can I use common handtools and/or test instruments to make the repairs?

If the answer is "no" to any of these questions, you simply are not ready to start troubleshooting any MTS equipment. Start studying!

2

MTS TV
STEREO GENERATOR

As discussed in Chapter 1, it is essential that a special-purpose generator be used to test and adjust the MTS decoder circuits (sometimes called the *demodulator* circuits, just to confuse you). This is because conventional TV generators (including the most advanced) do not produce the signals that simulate stereo and SAP broadcasts.

Even though there are few adjustment controls on an MTS decoder (typically three or four controls, or fewer), precise signals must be available to make the adjustments and to perform tests on the circuits. Because these signals are of such importance to the service technician, this entire chapter is devoted to an instrument that generates the required signals.

2-1. MODEL 2009 MTS TV STEREO GENERATOR

The instrument chosen is the B&K-Precision/Dynascan Model 2009 MTS TV Stereo Generator, shown in Fig. 2-1. The Model 2009 uses a system of *spot modulation* at frequencies of 300 Hz, 1 kHz, and 8 kHz. This provides all the test frequencies generally needed for full testing, servicing, troubleshooting, and adjustment of stereo-TV equipment.

The modulation levels at each of the "spots" are preset to *simulate dbx encoding.* This design approach makes the Model 2009 very cost-effective, rather than trying to duplicate the very complex and expensive dbx noise-reduction scheme across the entire audio band. (To duplicate the dbx noise reduction over the whole audio band requires an instrument similar to those found in broadcast studio work, costing thousands of dollars.)

FIGURE 2-1   Model 2009 MTS TV-Stereo Generator. (Courtesy of Dynascan Corporation/B&K-Precision.)

The 2009 generates an MTS FM-modulated radio-frequency (r-f) carrier signal on TV channel 3 or 4, and at the standard TV intermediate-frequency (i-f) frequency, as well as on a standard 4.5-MHz audio carrier. Thus there are carriers (with modulation) at all three TV frequencies (RF, IF, and audio). The modulating signal itself (composite audio) is also available for injection directly into audio and decoder circuits.

A mode-selector switch permits four combinations of internal modulation: (L) left channel only, (R) right channel only, (L + R) baseband, and (L − R) subband. These four combinations of modulation permit complete testing of the stereo decoder circuit, as well as channel balance and channel separation of the audio circuits that follow the decoders. As an added feature, video modulation accompanies audio modulation to give an indication on the TV screen of left channel only, right channel only, or both channels.

A highly stable 15.734-kHz pilot signal is generated and combined with the composite audio. The pilot signal may be switched off when desired. This makes it possible, quickly and effectively, to test the *pilot detector* circuits in the decoder.

A SAP mode is also selectable. This provides a subband signal centered at 78.67 kHz to test SAP operation.

For servicing convenience, the modulation levels are preset to provide an equal-level audio output signal at 300 Hz, 1 kHz, or 8 kHz (in a properly operating TV set). The modulation levels are also preset to provide equal audio output levels for each of L, R, (L + R), and (L − R) modulation modes. This permits quick checking of proper operation on the equipment being serviced.

2-2. MTS TV GENERATOR SPECIFICATIONS

The following specifications apply to the Model 2009 generator.

MAIN CHANNEL (MONAURAL)

Main-channel carrier frequency	4.5 MHz
Modulating signal	L + R
Internal modulating frequencies	300 Hz, 1 kHz, 8 kHz, 1 kHz at −10 dB
Frequency accuracy	±1%
Preemphasis	75 μs
External input	
Frequency range	100 Hz to 10 kHz
Input voltage range	2 V p-p maximum
Input impedance	10 kΩ

PILOT SUBCARRIER

Frequency	fH = 15.734 sine wave
	Pilot subcarrier frequency is synchronized to horizontal sync pulses of composite sync
Audio (4.5 MHz) carrier deviation by pilot subcarrier	±5 kHz

STEREOPHONIC SUBCHANNEL

Modulating signal	L − R
Subcarrier frequency	2 fH = 31.468 kHz
Subcarrier modulation method	AM-DSBSC (amplitude modulation, double-sideband suppressed carrier)
Audio (4.5 MHz) carrier deviation by modulated stereophonic subcarrier	±40 kHz maximum
Audio (4.5 MHz) carrier deviation by main-channel signal plus modulated stereophonic subcarrier	±50 kHz
Difference between time-axis crossings of pilot subcarrier and stereophonic subcarrier (in pilot-frequency degrees)	3° maximum
Modulating frequencies (internal only) used in dbx simulation	300 Hz, 1 kHz, 8 kHz, 1 kHz at −10 dB

COMPOSITE STEREO MODULATION

Stereophonic separation (300 Hz, L or R only)	25 dB minimum
Crosstalk of stereophonic subchannel into main channel	−40 dB nominal
Crosstalk of main channel into stereophonic subchannel	−40 dB nominal

STEREO AUDIO PROGRAM (SAP) SUBCHANNEL

Subcarrier frequency	5 fH = 78.67 kHz
Subcarrier frequency tolerance	±500 Hz
Subcarrier modulation method	FM

Modulating frequencies (internal only) used in dbx simulation	300 Hz, 1 kHz, 8 kHz
Subcarrier deviation (300 Hz, (L−R)	±2.4 kHz
Audio (4.5 MHz) carrier deviation by SAP subcarrier	±15 kHz maximum

AUDIO (4.5 MHz) CARRIER

Frequency	4.5 Mhz
Tolerance	±0.10 MHz
Audio modulation bandwidth	120 kHz
Deviation range	±100 kHz
Deviation monaural (300 Hz, L+R, −12 dB below 100% deviation)	±6.3 kHz

OUTPUT SIGNAL (RF)

RF output frequency

Channel 3	61.25-MHz video carrier 65.75-MHz sound carrier
Channel 4	67.25-MHz video carrier 71.75-MHz sound carrier
IF	45.75-MHz video carrier 41.25-MHz sound carrier
Modulation	Composite sync with right/left indicator
Output impedance	75-Ω unbalanced 300-Ω balanced with adapter
Output level	+20 dBmV

OUTPUT SIGNAL (COMPOSITE STEREO)

| Output impedance | 600 Ω |
| Output level (8 kHz, L or R only) | 0–2 V p-p |

MISCELLANEOUS

Input power	120 V ± 10%, 60 Hz, 10 W
Operating temperature	0 to +45 °C
Dimensions (H×D×W), including handle	3.4 × 11.5 × 9.2 in. (87 × 293 × 234 mm)
Weight	3.4 lb (1.5 kg)

2-3. MTS TV GENERATOR ACCESSORIES

The following accessories are available for the Model 2009 generator.

Detachable a-c power cord.
Instruction manual.
Schematic diagram and parts list.

Model CC-55 BNC-to-F Cable. This cable connects the BNC output connectors of the Model 2009 to type F connectors on the unit under test. The cable is a 5-ft length of 75-Ω RG-59/U coaxial cable with a BNC connector on one end, and a type F connector on the other.

Model CC-56 BNC-to-BNC Cable. Connects the BNC output connectors of the Model 2009 to other test equipment using BNC connectors. This cable is a 5-ft length of 75-Ω RG-59/U coaxial cable with BNC connectors on each end.

Probes. An oscilloscope probe may be used to inject signals from the Model 2009 into the unit under test. Probes with attenuation, such as the fixed 10:1 type, are not suitable. Use a nonattenuating type of probe, or use the DIRECT setting for probes that can be switched between 10:1 and direct.
The following B&K-Precision oscilloscope probes are recommended:

Model PR-40 10:1/direct probe
Model PR-37 Deluxe 10:1/direct probe

2-4. COMPOSITE OR TYPICAL DECODER CIRCUITS

The Model 2009 MTS TV Stereo Generator is designed to test, adjust, and trouble-shoot MTS decoder circuits, such as shown in Fig. 2-2. We describe a cross section of specific decoder circuits in Chapters 3 through 7. Before we get to such descriptions, it may be helpful to discuss typical or overall decoder functions (found in virtually all decoder circuits). In that way you can relate the Model 2009 signals and features to decoder functions. This also serves as a starting point to an understanding of the various refinements and additions to the basic decoder functions found in some stereo-TV/VCR circuits.
While the Model 2009 is similar to a low-power stereo-TV transmitter in many ways, the 2009 is simplified in some respects and more versatile in other respects. The Model 2009 is more versatile in that individual selection of each modulation mode (L + R), (L − R), L, R, pilot, or SAP is possible. You must take what you get when testing or adjusting using a stereo-TV broadcast. Also, a stereo-TV broadcast is available only in RF, whereas the 2009 provides RF, IF, 4.5-MHz audio carrier, and composite audio signals.

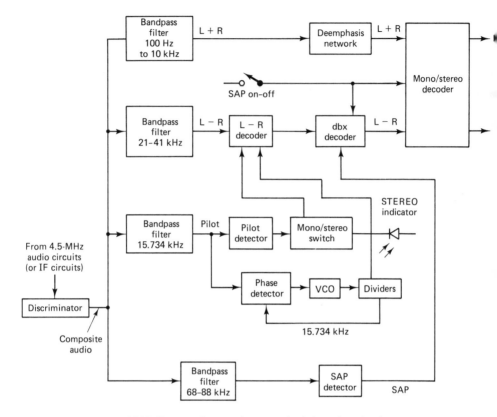

FIGURE 2-2   Composite or typical decoder circuits.

The Model 2009 is simplified by not providing dbx encoding for the entire audio spectrum. (The 2009 uses spot modulation at 300 Hz, 1 kHz, and 8 kHz). Two audio levels are provided at 1 kHz, and one audio level at 300 Hz and 8 kHz. The FM deviation is preset for each of these spots to duplicate that of dbx encoding. As a result, the 2009 simulates dbx encoding at sufficient test frequencies and levels to perform all necessary measurements and adjustments.

Keep the following points in mind when studying the circuits of Fig. 2-2. In many television receivers, the output of the 4.5-MHz audio section is applied to the discriminator. In other designs, the discriminator is fed directly from the sound IF circuits which are filtered to eliminate the picture IF. (That is why the 2009 provides RF, IF, and 4.5-MHz outputs, all with the same modulation capability.)

No matter what television receiver design is used, an MTS discriminator output is a composite audio signal similar to that shown in Fig. 2-3. This basic composite audio signal consists of the 50-Hz to 15-kHz baseband (L + R), 15.734-kHz pilot signal, 31.468-kHz stereo subchannel (L − R), and 78.670-kHz SAP subcarrier.

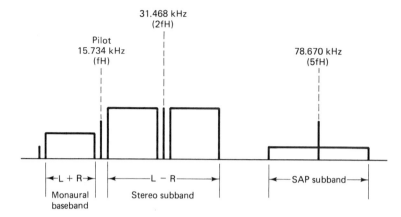

FIGURE 2-3 Typical composite audio signal input to decoder circuits.

2-4.1 VCO circuit

During stereo operation, a PLL locks the sine-wave output of a VCO in phase with the received 15.734-kHz pilot signal. The VCO operates at some multiple of the pilot signal and is divided down to 15.734 kHz.

2-4.2 Pilot detector and mono/stereo switch

A pilot detector circuit senses the presence of the pilot signal and operates the mono/ stereo switch. In turn, the mono/stereo switch enables the L–R decoder and operates the STEREO indicator (typically a front-panel LED). Note that the pilot signal can be turned on and off at the 2009. This makes it possible to check quickly the VCO, pilot detector, and mono/stereo switch circuits. During monaural operation, or during any loss of the pilot signal, phase lock cannot be maintained and the mono/stereo switch circuit disables the L–R decoder (as well as turning off the STEREO indicator).

2-4.3 L+R channel

Bandpass filters separate the L+R baseband and the L–R subchannel from the composite audio. Although the transmitted baseband signal may have a bandwidth of about 50 Hz to 15 kHz, with sharp suppression above 15 kHz, the typical receiver circuit may have a bandwidth that is closer to 100 Hz to 10 kHz.

The L+R signal is fed through a deemphasis network to recover the L+R signal. After deemphasis, the L+R signal has the same characteristics as before preemphasis at the broadcast transmitter.

2-4.4 L-R channel

The L−R signal is an AM, double-sideband suppressed-carrier signal centered about a 31.468-kHz subcarrier. The L−R bandpass filter typically has a bandwidth of about ±10 kHz on either side of the subcarrier (from about 21 to 41 kHz). This signal is applied to the L−R decoder circuit together with a 31.468-kHz carrier. This carrier is derived from the VCO and dividers, and is phase-locked to the pilot signal. (The L−R decoder is enabled only during stereo operation, when a pilot signal is present.)

The output of the L−R decoder is the L−R signal with the subcarrier removed. This output signal still contains the dbx encoding added by the transmitter.

The L−R signal is routed through a dbx decoder, which complements the dbx encoding of the transmitter to restore the L−R signal and takes advantage of the noise reduction possible through the dbx system.

The dbx decoder provides an L−R signal (only) to the mono/stereo decoder. In turn, the mono/stereo decoder splits this single output into identical L and R signals for the audio amplifiers.

2-4.5 Mono/stereo decoder

The L+R and L−R signals are combined in the mono/stereo decoder in such a manner to produce separate L and R audio signals. During monaural operation, the L+R signal is equally split into L and R outputs.

An alternative configuration to that shown in Fig. 2-2 may be a *frequency-division demultiplexer* or *time-division demultiplexer* which has a composite audio input and a 31.468-kHz input, with L+R and L−R outputs.

When L+R and L−R signals are applied to the inputs of the mono/stereo decoder, the decoder separates the L and R components into independent left- and right-channel outputs. When the L−R input is absent, such as during monaural reception, the L+R input is separated into equal L and R components.

2-4.6 SAP channel

When the composite audio (Fig. 2-3) includes an SAP signal centered about 78.670 kHz, a bandpass filter with a bandwidth of about ±10 kHz (from about 68 to 88 kHz) passes the SAP signal and blocks other components of the composite audio.

The SAP signal is recovered by the FM SAP detector. At this point, the SAP signal still has the dbx characteristics that are added at the transmitter. When SAP is selected (typically by a front-panel control), both mono and stereo reception are disabled, and the SAP signal is applied to the dbx decoder in place of mono (L+R) and stereo (L−R). The dbx decoder restores the original audio characteristics to the SAP signal.

2–5. MTS TV GENERATOR OPERATING CONTROLS AND INDICATORS

The following descriptions apply to the Model 2009 generator shown in Fig. 2–1.

COMPOSITE LEVEL Control. Adjusts level available at COMPOSITE output jack only. This control does not affect the internal or external modulation levels when using 4.5-MHz or IF/RF outputs.

ON Indicator. Turns on when instrument is operating.

POWER Switch. Turns instrument on and off.

SAP (Second Audio Program) Switch. Selects monaural SAP using dbx simulation. Stereo operation is disabled regardless of PILOT switch position. (Normally, L−R modulation is used.) Spot modulation of the SAP signal at 300 Hz, 1 kHz, or 8 kHz is selectable with the AUDIO FREQ (Hz) switches.

PILOT Switch. Turns 15.734-kHz pilot subcarrier on and off. Must be on to activate the stereo function in the decoder under test.

AUDIO FREQ (Hz) Switches. Four interlocking pushbutton switches select internal modulation frequencies of 300 Hz, 1 kHz, 8 kHz, or 1 kHz at −10 dB. Each modulating frequency is preadjusted to simulate the propr dbx amplitude and phase.

MODULATING SIGNAL Switches. Five interlocking pushbutton switches select the type of modulating signal.

OFF, EXT: No internal sound modulation. Signal applied at EXT AUDIO INPUT jack becomes the COMPOSITE output and monaural L+R-type modulation of 4.5-MHz and IF/RF ouptuts. IF/RF output is also modulated with composite sync, but neither left or right visual indicators are produced.

L: Internal modulation by left-channel signal only at frequency selected by AUDIO FREQ (Hz) switches. Left-channel portion of L+R baseband and L−R signal at COMPOSITE output. Left-channel modulation of 4.5-MHz and IF/RF outputs. IF/RF output is also modulated by composite sync with left visual indicator.

R: Internal modulation by right-channel signal only at frequency selected by AUDIO FREQ (Hz) switches. Right-channel portion of L+R baseband and L−R signal at COMPOSITE output. Right-channel modulation of 4.5-MHz and IF/RF outputs. IF/RF output is also modulated by composited sync with left visual indicator.

L+R: Left- and right-channel signals are in phase. Internal modulation by L+R signal at frequency selected by AUDIO FREQ switches. L+R baseband (only) present at COMPOSITE OUTPUT. L+R modulation of 4.5-MHz and IF/RF outputs.

IF/RF output is also modulated by composite sync with both left and right visual indicators.

L−R: Left- and right-channel signals are out of phase. Internal modulation by L−R signal at frequency selected by AUDIO FREQ (Hz) switches. L−R signal (only) present at COMPOSITE output. L−R modulation of 4.5-MHz and IF/RF output is also modulated by composite sync with both left and right visual indicators.

IF Switch. When the IF switch is engaged (pushed in) an IF carrier frequency output (45.75 MHz) is selected at the IF/RF jack. When the IF switch is released (out), the RF switch determines the carrier frequency at the IF/RF jack.

RF Switch. When the IF switch is released, the RF switch selects the carrier frequency at the IF/RF output jack. CHAN 4 (67.25 MHz) is selected when the RF switch is engaged (pushed in). CHAN 3 (61.25 MHz) is selected when released (out).

IF/RF Output Jack. Provides I-F or R-F carrier output at the frequency selected by the IF and RF switches. This signal is FM modulated by composite audio, and AM modulated by composite sync with left/right visual indicators.

4.5-MHz Output Jack. Provides 4.5-MHz audio carrier. This signal is FM modulated by composite audio.

COMPOSITE Output Jack. Provides composite of the selected MODULATING SIGNAL and/or PILOT. Provides SAP when selected. Output level is adjustable with COMPOSITE LEVEL control.

EXT AUDIO INPUT Jack. Accepts external monaural audio input when MODULATING SIGNAL switch is set to OFF, EXT. Provides L+R type modulation. Accepts the frequency range 100 Hz to 10 kHz. About 2.25 V p-p input provides 100% deviation at 300 Hz; about 0.58 V p-p input provides 100% deviation at 8 kHz.

2–6.   MTS TV GENERATOR OPERATING PROCEDURES

The following procedures apply to the Model 2009 generator when used to adjust, test, and troubleshoot stereo-TV receivers, stereo adapters, VCRs with stereo capability (hi-fi VCRs), and so on.

2–6.1   *Output signals*

Four types of output signals are available for injection at various points in the equipment under test.

1. The RF/IF output signal may be injected at the antenna terminals (of a TV or VCR) on channel 3 or channel 4 when the IF switch is released.

2. The RF/IF output signal may be injected into the i-f section (of a TV or VCR) when the IF switch is engaged.

3. The 4.5-MHZ output signal may be injected directly into the 4.5-MHz audio circuits.

4. The COMPOSITE output signal may be injected directly into various points in the stereo decoder circuits.

As discussed in Sec. 2–5, switches permit selection of L, R, L + R, L − R, pilot, or SAP signals at test frequencies of 300 Hz, 1 kHz, or 8 kHz.

2–6.2 Initial setup

1. Connect the power cord of the Model 2009 generator to an a-c outlet.

2. Turn on the 2009 by pressing the POWER switch (Fig. 2–1). The ON indicator should turn on.

3. Apply power to the decoder circuits or unit under test (stereo TV, hi-fi VCR, etc.). If the equipment being tested has a "hot chassis," use an isolation transformer. (The author recommends an isolation transformer for all tests, where practical.)

2–6.3 RF output

An r-f output signal, modulated with mono, stereo, or SAP audio, may be injected at the antenna terminals of the unit under test (TV or VCR) on VHF channel 3 or 4, if desired. This provides an antenna-to-loudspeaker check of the audio path, including the RF tuner, IF amplifier, stereo decoders, and audio amplifiers.

The r-f output is continuous and may be used at the same time as the 4.5-MHz and COMPOSITE outputs, if desired. The RF signal level is fixed at an appropriate level for injection at the antenna terminals.

The r-f output signal also contains *composite sync* (to lock the video circuits) and a *visual indicator* which displays a *rectangle* on the video screen. This provides an antenna-to-screen check of the video path, including RF tuner, IF amplifier, and video circuits.

The rectangle appears on the left half of the screen when L (left) modulation is selected, and vice versa when R (right) modulation is used. A rectangle appears on both halves of the screen when L + R or L − R modulation is selected.

Proceed as follows to use the r-f output.

1. Connect a 75-Ω coaxial cable (RF-59/U) from the IF/RF output jack (Fig. 2–1) to the antenna terminals of the TV or VCR. The 75-Ω type F VHF input is preferred, or use a VHF 75-to-300 coupler.

2. Set the channel selector of the TV or VCR to channel 3 or channel 4, whichever is normally unused in the broadcast area.

3. Release the IF switch (pull out), and select CHAN 3 or CHAN 4 with the RF switch (channel 4 in, channel 3 out). Make certain that both the TV/VCR and 2009 are tuned to the same channel!

4. The 2009 is now ready to perform both stereo and SAP tests through the r-f circuits.

2–6.4 IF output

An i-f output signal, modulated with mono, stereo, or SAP audio, may be injected into the i-f section of the TV or VCR on the standard i-f frequency of 45.75 MHz, if desired. This provides an IF-to-loudspeaker check of the audio path, including the i-f amplifier, stereo decoders, and audio amplifiers.

The i-f output is continuous and may be used at the same time as the 4.5-MHz and COMPOSITE outputs, if desired. The i-f signal level is fixed at an appropriate level for injection at the i-f amplifier section.

The i-f output signal also contains composite sync (to lock the video circuits) and a visual indicator which displays a rectangle on the video screen. This provides an IF-to-screen check of the video path, including IF amplifier and video circuits. The rectangle appears on the left, right, or both halves of the screen, as described for r-f output in Sec. 2–6.3.

Proceed as follows to use the i-f output.

1. Connect a probe to the IF/RF ouptut jack (Fig. 2–1).
2. Press the IF button.
3. The probe may be used to inject the 45.75-MHz i-f signal at the desired point.
4. The 2009 is not ready to perform both stereo and SAP tests through the i-f circuits, up to the stereo decoder circuits.

2–6.5 4.5-MHz output

A 4.5-MHz output signal, modulated with mono, stereo, or SAP audio, may be injected into the 4.5-MHz section of the TV or VCR, if desired. This provides a 4.5-to-loudspeaker check of the audio path, including the stereo decoders and audio amplifiers.

The 4.5-MHz output is continuous and may be used at the same time as the IF/RF and COMPOSITE outputs, if desired. The 4.5-MHz signal level is fixed at an appropriate level for injection at the 4.5-MHz circuits.

Proceed as follows to use the 4.5-MHz output.

1. Connect a probe to the 4.5-MHz output jack (Fig. 2–1).
2. The 4.5-MHz output is continuous, so the probe may be used to inject the 4.5-MHz signal, no matter what operating mode is selected.

3. The 2009 is now ready to perform both stereo and SAP tests through the 4.5-MHz circuits, including the stereo decoder.

2–6.6 *COMPOSITE output*

A composite output signal, modulated with mono, stereo, or SAP audio, may be injected into the stereo decoder circuits of the TV or VCR, if desired. This provides a decoder-to-loudspeaker check of the audio path, including the stereo decoders and audio amplifiers. The COMPOSITE output is continuous, but variable, and may be used at the same time as all other operating modes.

The SAP, PILOT, AUDIO FREQ (Hz), and MODULATING SIGNAL switches permit selection of L, R, L + R, L − R, pilot, or SAP signals to match the type of signal normally present at each point in the decoder. The signal level is set by the COMPOSITE LEVEL control.

Proceed as follows to use the COMPOSITE output.

1. Connect a probe to the COMPOSITE output jack (Fig. 2–1).
2. The COMPOSITE output is continuous, so the probe may be used to inject the composite output signal, no matter what operating mode is selected. However, keep in mind that the signal level (from about 0 to 2 V) is set by the COMPOSITE LEVEL control.
3. The 2009 is now ready to perform both stereo and SAP tests through the stereo decoder circuits.

2–6.7 *Stereo tests*

Stereo tests are generally conducted at the r-f, i-f, or 4.5-MHz level (although it is possible to make stereo tests on some decoders at the composite-signal level). In Chapters 3 through 7 we discuss many specific stereo tests for a particular decoder. However, virtually all decoders require a *stereo separation* test and a *stereo indicator* (pilot) test. The basic procedures are as follows.

1. For all stereo tests, the PILOT button (Fig. 2–1) must be pressed. If the PILOT button is out, there is no pilot signal and all stereo functions are disabled (the decoder circuits operate in mono and the STEREO indicator goes out).

2. Before making any stereo tests, make certain that the stereo function is selected on the unit under test. Typically, this means setting a STEREO/MONO switch to STEREO, or an AUTO/MONO switch to AUTO, or possibly setting a STEREO/SAP switch to STEREO. Always check the operating instructions for the unit under test.

3. Select the desired audio modulating frequency of 300 Hz, 1 kHz, or 8 kHz, using the appropriate AUDIO FREQ (Hz) switch. You can also use the 1 kHz–10 dB switch, but this results in a reduction in output voltage (of about 3:1).

4. Select the desired form of modulation L, R, L − R, or L + R, using the appropriate MODULATING SIGNAL switch. Many stereo tests are conducted using

L (left), followed by a repeat of the test with R (right) modulation. Often, this is followed with a test using L − R modulation.

Keep in mind that if you select L + R modulation, there should be no stereo operation at the decoder (even though the STEREO indicator on the unit under test should remain on if the PILOT button is in). In most decoders, the STEREO indicator remains on, but you get mono reception if both L + R and PILOT are in.

Also keep in mind that the MODULATING SIGNAL switches are interlocked, so you can select only one of the four modulation signals at a time (L, R, L + R, or L − R). The PILOT switch is independent of the MODULATING SIGNAL switches.

5.   *To measure stereo separtion,* select L (with the PILOT switch in). The STEREO indicator on the unit under test should turn on, sound should be heard from the left speaker, and a rectangle should be displayed on the left half of the screen (if you are using either r-f or i-f outputs, but not with 4.5-MHz or COMPOSITE outputs).

6.   Connect an audio voltmeter to the left-channel loudspeaker terminals (or at the left-channel output, whichever is convenient). Measure the left-channel output voltage.

7.   Select R (with the PILOT switch in). The STEREO indicator on the unit under test should remain on, should be heard from the right speaker, and the rectangle should be displayed on the right half of the screen (with either r-f or i-f outputs).

8.   Measure the right-channel output voltage and compare the right-channel voltage to the left-channel voltage measured in step 6.

The ratio of the left-channel output to that of the right-channel output is the *stereo separation* and is usually expressed in decibels. Figure 2–4 can be used to convert voltage ratios to decibels. For example, a measured voltage ratio of 10:1 corresponds to 20 dB. The actual voltages measured are of no particular importance, it is the *voltage ratio* that counts. However, both voltages (L and R) must be measured at the same point in the audio path (typically at the L and R loudspeakers), and both must be measured using the same input voltage level and frequency.

An audio voltmeter such as the B&K-Precision Model 295 or 297 Wideband AC Voltmeter is ideal for stereo separation measurements. The 295 and 297 read out in both decibels and voltage. The 297 is a dual-input instrument with a dual-point meter. This permits both readings to be compared simultaneously.

9.   *To check the STEREO indicator* of the unit under test, press and release the PILOT button when any one of the MODULATING SIGNAL buttons are in. The STEREO indicator should turn on and off.

Note that when the PILOT button is in (STEREO indicator on), the channel outputs may (or may not) be equal. However, with the PILOT button out (STEREO indicator off), the channel outputs should be substantially the same, no matter what MODULATING SIGNAL button is pressed. In virtually all decoder circuits, the unit under test (TV or VCR) reverts to monaural operation when there is no pilot carrier broadcast (or when the simulated pilot carrier is removed).

Voltage Ratio	dB	Voltage Ratio	dB
1:1	0	11:1	21
1.1:1	1	12.5:1	22
1.25:1	2	14:1	23
1.4:1	3	16:1	24
1.6:1	4	18:1	25
1.8:1	5	20:1	26
2:1	6	22.5:1	27
2.25:1	7	25:1	28
2.5:1	8	28:1	29
2.8:1	9	31.5:1	30
3.15:1	10	35.5:1	31
3.55:1	11	40:1	32
4:1	12	45:1	33
4.5:1	13	50:1	34
5.0:1	14	56:1	35
5.6:1	15	63:1	36
6.3:1	16	71:1	37
7.1:1	17	80:1	38
8.0:1	18	90:1	39
9.0:1	19	100:1	40
10:1	20		

FIGURE 2-4 dB/voltage-ratio conversion.

2-6.8 SAP tests

SAP tests are generally conducted at the r-f, i-f, or 4.5-MHz level (although it is possible to make SAP tests on some decoders at the composite-signal level). We discuss many specific SAP tests in Chapters 3 through 7 for a particular decoder. However, virtually all SAP tests follow a certain pattern, which we discuss in the following steps.

Keep in mind that the SAP test signal is monaural, but with dbx encoding. Also, on the Model 2009, the output is monaural when the SAP button is pressed, no matter what the position of the PILOT button. This is not necessarily the case of a TV station signal where both stereo and SAP may be broadcast simultaneously (even though you can listen to only one at a time).

Proceed as follows to perform the basic SAP test.

1. Press the SAP button and release the PILOT button.

2. Before making any SAP tests, make certain that the SAP function is selected on the unit under test. Typically, this means setting a STEREO/SAP switch to SAP. Always check the operating instructions for the unit under test.

3. Select the desired audio modulating frequency of 300 Hz, 1 kHz, or 8 kHz,

using the appropriate AUDIO FREQ (Hz) switch. You can also use the 1 kHz–10 dB switch, but this results in a reduction in output voltage (of about 3:1).

4. Select the desired form of modulation L, R, or L–R, but not L + R. (When L + R is selected, together with SAP, there is no output.)

Generally, L–R is selected for SAP tests. This is because the L or R modes produce a lower output than L–R, since only that portion which becomes an L–R signal produces SAP modulation. No matter which form of modulation is selected, the output is monaural.

5. Check that the SAP indicator turns on and that the audio voltages at both loudspeakers are approximately equal (indicating monaural operation).

In some decoders it may be helpful to check SAP operation at L, then at R, and then at L–R. The L and R output indications (loudspeaker voltages) should be approximately equal, while the L–R output should produce a higher (but equal) voltage at both loudspeakers.

2–6.9   *Audio output-level (or separation) adjustment*

Most stereo decoder circuits require adjustment of the audio output level. (This is called the *separation* adjustment in some literature.) No matter what it is called, the adjustment is critical to establish the proper reference level into the dbx circuits (or from the dbx circuits into the audio amplifiers). Either way, improper adjustment usually degrades stereo performance to the point where separation approaches that of monaural operation.

Always use the audio output level (separation) adjustment procedures specified in the service literature for the unit under test. Some procedures require adjustment to obtain a specific audio level at the FM detector output, while applying a signal that is modulated with a 300-Hz L + R signal at 100% modulation. (Note that is some circuits the adjustment is in the L–R audio path, and is measured after the matrix that combines the L + R and L–R signals.)

The output of the Model 2009 is 12 dB below 100% deviation at 300 Hz when L + R is selected. This means that you should adjust the audio level for 12 dB below the FM-detector output level specified in the service literature (but using the same adjustment procedure). As shown in Fig. 2–4, a 12-dB reduction in level involves a 4:1 ratio, so use an FM-detector output level that is about 25% of that specified in the service literature.

An alternative method of adjustment is to select the *external modulation* mode of the 2009 (Sec. 2–6.10) and apply a 300-Hz external modulating signal. (An external modulating signal level of 2.26 V p-p (800 mV rms) at 300 Hz provides 100% deviation.)

2–6.10   *External modulation*

The output signal of the Model 2009 may be modulated externally as follows:

1. Press the OFF/EXT MODULATING SIGNAL button.
2. Connect the desired audio source to the EXT AUDIO INPUT jack. This signal must be in the range 100 Hz to 10 khz.
3. The external audio input level should not exceed 100% deviation. A voltage level of about 2.26 V p-p (800 mV rms) provides 100% deviation at 300 Hz. About 0.58 V p-p (205 mV rms) provides 100% deviation at 8 kHz. If voice or music signals are used, peaks should not exceed about 200 mV rms.

2–6.11 Operating notes

We provide specific procedures for use of the Model 2009 in Chapters 3 through 7. The following notes apply to all procedures and equipment under test.

The Model 2009 is designed and adjusted to provide the *same audio output level* in the unit under test at modulating frequencies of 300 Hz, 1 kHz, and 8 kHz. When the 1 kHz–10 dB modulation is selected, the audio output level in the unit under test should be 10 dB lower. This information can be useful in checking operation of the unit under test (to form a basis for troubleshooting).

The L, R, L+R, and L−R modes should produce equal audio output in the unit under test. That is, the audio levels in the left and right channels should be equal for all modulation modes. The L+R or L−R mode should produce equal audio level in each channel. Similarly, the L or R mode should produce the same level in the active channel as in the L+R or L−R modes. Again, this information can be used in checking operation to form a basis for troubleshooting.

For example, under a given set of conditions (same measurement points, same audio frequencies) the L and R channels should produce the same audio output voltage. If not, an unbalance condition is indicated. (This unbalance condition is not necessarily limited to the stereo decoder circuits but could as likely be in the audio amplifier circuits.)

The SAP function should produce equal SAP audio output levels at 300 Hz, 1 kHz, and 8 kHz. The SAP function is always monaural and should normally be checked only with the L−R modulation mode. As discussed, L or R produce lower output levels, whereas L+R produces no output, when SAP is used.

3

MITSUBISHI
MTS TV CIRCUITS

This chapter is devoted to the decoder circuits used in typical Mitsubishi stereo-ready TV sets. Note that the circuits are referred to as MCS, or multichannel sound, in Mitsubishi TV literature. The PC board containing the stereo-TV circuits is labeled the MCS PCB.

The decoder circuits receive their input in the form of composite audio from a sound i-f (SIF) detector. The broadcast r-f signal is converted to an i-f signal by a tuner located on the main PB board. The i-f signal is converted to a 4.5-MHz sound carrier by an i-f IC, also located on the main board.

The composite audio is extracted from the 4.5-MHz carrier by the SIF detector IC located on a sound PC board. The composite audio is amplified on the sound board and applied to the input of the decoder circuits on the MCS board. The output of the decoder circuits is applied to dual speakers (mounted on each side of the TV set) through a stereo amplifier (also located on the MCS board).

3–1.   OPERATING (USER) CONTROLS AND INDICATORS

Two front-panel operating (or user) controls and four front-panel indicators are associated with the decoder circuits. Figure 3–1 shows a typical arrangement for the controls and indicators, as well as the truth table.

The STEREO/SAP switch permits the user to select either STEREO or SAP operation. The AUTO/MANUAL switch determines the conditions under which STEREO or SAP is selected (which signal is fed to the audio amplifiers).

With the AUTO/MANUAL switch in AUTO and the STEREO/SAP switch in STEREO, the audio amplifiers (and loudspeakers) are connected for stereo (no matter which signal, stereo or SAP, is being broadcast).

With the AUTO/MANUAL switch in AUTO, and the STEREO/SAP switch in SAP, the audio amplifiers are connected for SAP. However, if the SAP broadcast should stop, the circuits switch to stereo automatically.

With the AUTO/MANUAL switch in MANUAL and the STEREO/SAP switch in STEREO, the circuits are connected for stereo (no matter which signal is being broadcast). With the AUTO/MANUAL switch in MANUAL and the STEREO/SAP switch in SAP, the circuits are connected for SAP. If there is no SAP broadcast, there will be no sound, even though a stereo broadcast may be present on the selected channel. Keep this in mind when troubleshooting a "the set went dead when I pushed the SAP button" symptom!

Also note that there are two front-panel LEDs that indicate the presence (or absence) of stereo and SAP broadcasts. The BROADCAST STEREO and BROADCAST SAP indicators turn on when the corresponding signals are broadcast on the selected channel.

These broadcast indicators are not to be confused with two front-panel LEDs that indicate which operating mode is selected. The STEREO and SAP indicators turn on when the corresponding mode is selected by the STEREO/SAP switch (or switches, in some cases).

Keep this in mind when troubleshooting a "I pushed the SAP button, and the SAP light turned on, but I still can't hear a SAP broadcast" symptom. Patiently

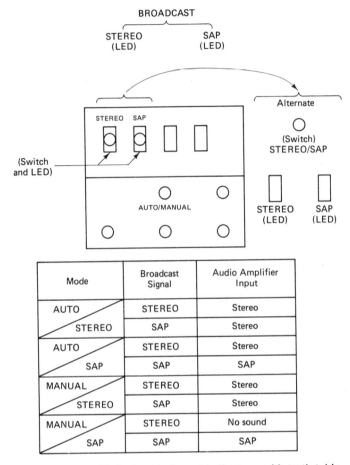

FIGURE 3–1   Typical controls and indicators, with truth table.

explain to the user that the BROADCAST SAP indicator must also be on, indicating that there is an SAP broadcast.

Also note that Fig. 3-1 shows two configurations for the STEREO/SAP switch (a single switch with two positions and two indicators, or two separate switches and indicators).

3–2. CIRCUIT DESCRIPTIONS

All of the notes described in Sec. 1-7 apply to the following descriptions.

3–2.1 Overall stereo decoder functions

Figure 3–2 is a block diagram showing the overall stereo decoder functions. We discuss the circuit details for two decoders using the overall configuration shown in Fig. 3–2. Before we get into these details, let us review the basic functions involved.

Signal separation for the decoder circuits is done by a network of bandpass and low-pass filters (BPFs and LPFs). These filters separate specific elements of the composite stereo signal for subsequent processing. Each element in the composite stereo signal can then be analyzed separately.

L + R. The composite stereo signal is applied to an LPF designed to pass only those frequencies in the L + R signal range (typically 0 to 15 kHz) and to block all other frequencies. As a result, only the L + R signal is presented to the L + R amplification circuits and to the *matrix circuit* to form the left and right stereo signals.

L − R. The sidebands comprising the L − R signal are separated from the composite stereo signal by a BPF (BPF-1) designed to pass frequencies below 46.5 kHz. This includes the 15.734-kHz pilot signal. As discussed in Chapter 1, the pilot is necessary to reinsert the 31.5-kHz subcarrier (suppressed at the transmitter) required to extract intelligence from the L − R signal during the decoding process.

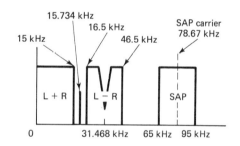

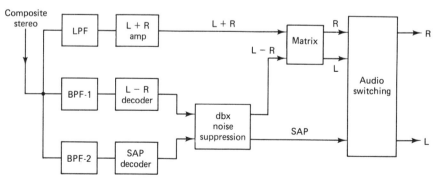

FIGURE 3–2 Overall stereo decoder functions.

The decoded L−R signal (typically in the range from 50 Hz to 15 kHz) is enhanced by a dbx noise-suppression process (restored to the original audio form) prior to application to the matrix circuit.

SAP. The SAP signal is isolated by a second bandpass filter (BPF-2) which passes only those frequencies within the range 65 to 96 kHz (the SAP subcarrier and associated sidebands).

The decoded SAP signal (extending from about 50 Hz to 12 kHz) is also subjected to noise suppression. After such suppression, the SAP signal is applied to an audio switch. This switch applies selected audio signals to the right and left channels of the stereo amplifier.

3–2.2 Stereo decoder signal path

Figure 3-3 shows the signal path of a typical stereo decoder in simplified form. The actual decoding process, including development of the reinserted 31.68-kHz subcarrier, is performed by a single IC (IC3A1). In Sec. 3–2.3 we describe how this is done. For now, simply consider that the decoded L−R signal (50 Hz to 15 kHz) is available at pin 5 of IC3A1.

Analog switch IC801 accepts either of two inputs applied to the dbx circuits. One input is the decoder L−R signal at pin 7, while the other input is the decoded SAP signal at pin 2. The choice between the two inputs depends on the logic signal at pin 3 of IC801. The logic, derived from Q803, is low at pin 3 of IC801 when stereo is selected, and high when SAP is selected. This logic signal is discussed further in Sec. 3–2.4.

Since the L−R and SAP signals are subjected to dbx processing at transmission, both signals must be restored to the original audio form before application to the audio amplifiers. This is done in the dbx IC802 (using a process such as that described in Chapter 1). After the noise-reduction process, the L−R and SAP signals appear at pin 8 of IC802 and are applied to the audio input switch through amplifiers and matrix circuits.

If the TV station is broadcasting in mono only (no stereo, no SAP), the L−R decoder is turned off (no pilot signal) and there is no L−R available. In addition, the SAP signal is disabled (no SAP carrier). Under these conditions, only the mono audio passes through LPF3A4, Q3A1, LPF3A1, Q3A3, Q806, and IC804 to the audio matrix (at the junction of R3D6/R3D7).

Note that the combined L−R and SAP signals at pin 4 of IC801 are applied to three inputs of dbx IC802, through BPFs and amplifiers Q804/Q805. The total L−R and SAP signal spectrum is applied to pin 18. Frequencies in the range 100 Hz to 3 kHz are applied to pin 3. Frequencies in the range 4 to 15 kHz are applied to pin 20.

Also note that the L−R signal is amplified by separate L−R amplifier IC803 and that the level of the L−R signal (after dbx processing) is set by VR3A5. The setting of VR3A5 is critical to proper operation of the decoder circuits. For example,

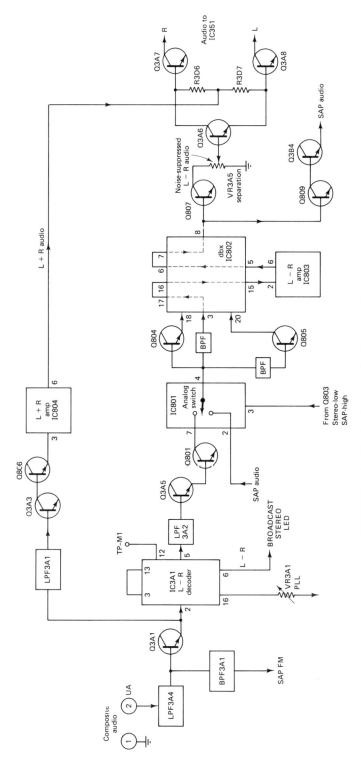

FIGURE 3–3 Basic signal path of a typical stereo decoder.

if VR3A5 is improperly set, you will hear only the mono audio or there will be no separation of left and right audio signals (so that stereo appears as mono). That is why VR3A5 (or the equivalent control in other stereo decoder circuits) is often called the *separation* adjustment control.

The mono and stereo signals are combined in the matrix circuit before application to the audio input switch IC351 (Sec. 3-2.5). Transistor Q3A6 functions as a phase splitter for the matrix network. Q3A6 has two outputs: (1) an inverted L−R signal at the collector for application to Q3A7, and (2) a noninverted L−R signal at the emitter (applied to Q3A8).

The L+R (mono) signal from pin 6 of IC804 is applied to the center tap of a voltage divider (consisting of R3D6/R3D7). The net result is the mixing, or matrixing, of the inverted and noninverted L−R signals with the mono L+R signal across R3D6/R3D7. The matrix produces signal inputs to the bases of Q3A7 and Q3A8 as follows.

The L+R signal is added algebraically across R3D6 (and applied to Q3A7) to the inverted L−R signal, in accordance with the equation (L+R)+(inverted L−R)=2R+L−L=2R. As a result, the net signal applied to Q3A7 and available at the ouptut is the right-channel audio signal.

The L+R signal is added algebraically across R3D7 (and applied to Q3A8) to the noninverted L−R signal, in accordance with the equation (L+R)+ (noninverted L−R)=2L+R−R=2L. As a result, the net signal applied to Q3A8 and available at the output is the left-channel audio signal.

At this point in the audio path, the transmitted stereo audio signal (L−R) has been properly reconstructed for application to the left- and right-channel audio amplifiers through the audio input switch IC351.

Note that the SAP signal, at the output pin 8 of dbx IC802, does not require matrixing. Instead, the SAP signal is applied to IC351 through amplifiers Q809 and Q3B4.

3-2.3 L−R decoder

Figure 3-4 shows the L−R decoder circuits in simplified form. As discussed in Chapter 1, the L−R portion of the composite stereo signal is formed by amplitude-modulating the 31.468-kHz subcarrier, suppressing the subcarrier at the point of transmission, and transmitting only the AM sidebands. To extract the audio intelligence (contained in the sidebands) at the TV receiver, the 31.468-kHz subcarrier is restored.

L−R decoding is essentially a two-step process: (1) the development of a locally generated frequency- and phase-corrected 31.468-kHz subcarrier signal; and (2) use of the reproduced subcarrier to reconstruct the L−R audio signal from the transmitted sidebands.

For proper decoding, the locally generated 31.468-kHz subcarrier must be identical, in both frequency and phase, to that of the subcarrier suppressed during transmission. As a result, some form of *frequency reference* must be provided as

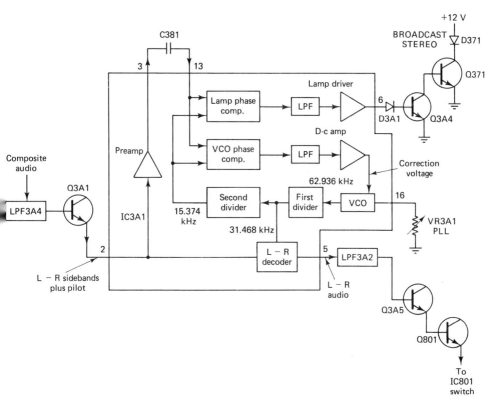

FIGURE 3–4 L–R decoder circuits.

a standard against which the locally generated signal can be compared. The 15.734-kHz pilot is used as the reference.

Actual frequency and phase comparison is provided by a conventional PLL, using phase comparators, and a VCO, all contained with IC3A1 as shown in Fig. 3–4. The fundamental operating frequency of the VCO is held to a frequency equivalent to *four times* that of the pilot (or 62.936 kHz). The output is automatically divided by 2 to get a 15.734-kHz sample of the VCO output. This divided sample is compared to the broadcast pilot signal to assure the accuracy and frequency stability of the VCO.

The L–R signal is applied to pin 2 of IC3A1 from the emitter of Q3A1. At this point in the path, the signal consists of the 15.734-kHz pilot, together with the L–R amplitude-modulated sidebands (without a carrier). The L–R signal is applied (unaltered) to the L–R *decoder* in IC3A1. The L–R signal is also applied to the inputs of *dual phase comparators* through a *preamp* stage and external capacitor C381.

The output of the VCO in IC3A1, after division by 2, is applied to the L–R decoder (at a frequency of 31.468 kHz) as a substitute carrier. After insertion of the substitute carrier, the decoder output at pin 5 of IC3A1 is an L–R audio signal, extending from about 50 to 15 kHz. The L–R signal is then subjected to dbx processing and amplification, as described in Sec. 3–2.2. LPF3A2 removes any frequencies above 15 kHz.

The dual phase comparators (*lamp phase comparator* and VCO *phase comparator*) also receive a 15.374-kHz signal, divided down twice from the 62.936-kHz VCO. The *VCO phase comparator* compares the VCO sample with the 15.374-kHz pilot signal and develops a correction voltage. The amplitude of the correction voltage depends on the degree of frequency or phase error detected between the sample and pilot signals. The resultant voltage is filtered, amplified, and applied to the VCO as a correction voltage. The correction voltage alters the fundamental operation frequency and phase of the VCO until the sample and pilot signals are identical. When this occurs, the loop is "locked" and the VCO is maintained at a frequency precisely four times that of the 15.734-kHz pilot signal (at 62.936 kHz).

Note that the VCO operating frequency can also be adjusted manually by PLL adjust VR3A1. As discussed in Sec. 3–3, the basic adjustment procedure is to monitor the VCO output (after division) with a frequency counter and adjust VR3A1 until the VCO output is correct.

The *lamp phase comparator* detects the presence of the pilot signal (indicating a stereo broadcast) and develops the output logic required to turn on the BROAD-CAST STEREO indicator LED D371. The lamp phase comparator receives both a sample 15.374-kHz signal (divided down frm the VCO) and a pilot signal (available only during a stereo broadcast). When both signals are present, pin 6 of IC3A1 goes low. This low is inverted to a high by Q3A4, driving Q371 into conduction, to turn D371 on.

3–2.4 SAP circuits

Figure 3–5 shows the SAP decoder circuits in simplified form. As discussed in Chapter 1, SAP is an FM signal, modulating a 78.6-kHz subcarrier, with sidebands extending from 65 to 95 kHz. Since the SAP signal is well above the highest frequency present in the composite stereo signal (46.5 kHz), the SAP signal may readily be separated from the composite signal by BPF3A1.

The SAP circuits may be divided into two major sections. One section processes the SAP signal and generates the SAP audio (which is amplified and applied to the loudspeakers). The other section senses the presence of SAP and generates commands that turn on the BROADCAST SAP indicator. These same commands are also used by the audio input select circuits.

As shown in Fig. 3–5, the FM SAP signal from BPF3A1 is applied to the input of an FM limiter within IC3A2. The limited FM signal is applied to a discrete-component FM demodulator through emitter follower Q3D1. The FM demodulator is comprised of T3A1, D3A2, and D3A3. The demodulated SAP audio is taken from the junction of D3A2/D3A3.

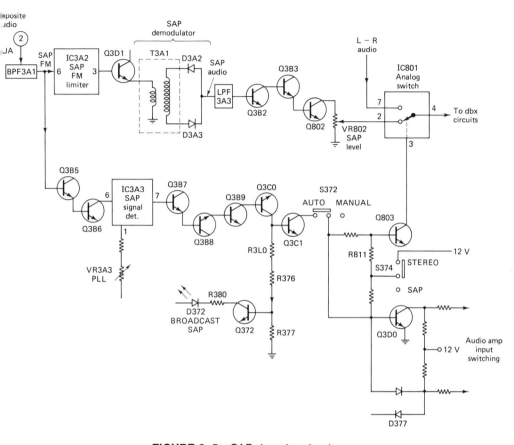

FIGURE 3-5 SAP decoder circuits.

The SAP audio is passed through LPF3A3 to remove any remainig high-frequency FM signal, and is amplified by Q3B2, Q3B3, and Q802. The amplified SAP signal is applied to pin 2 of IC801 through SAP level control VR802 (which sets the level of SAP signal in relation to the stereo and/or mono signal).

If the user has selected the SAP function, pin 3 of IC801 is high. This connects pin 4 of IC801 to pin 2. The SAP audio at pin 4 is then processed by the dbx circuits as described in Sec. 3-2.2. (If the user selects stereo instead of SAP, pin 3 of IC801 is low, and pin 4 of IC801 is connected to pin 7.)

The SAP signal from BPF3A1 is also applied to pin 6 of IC3A3 through amplifiers Q3B5 and Q3B6. IC3A3 is a conventional PLL detector that generates a low at pin 7 if a SAP signal is present. The low at pin 7 of IC3A3 is amplified by Q3B7, Q3B8, Q3B9, and Q3C0, and appears as a high at the collector of Q3C0. This high is applied to the base of Q372 through R3L0 and R376. The high drives Q372 into conduction, turning the BROADCAST SAP indicator LED D372 on, informing the user that a SAP signal is being broadcast.

The high from the collector of Q3C0 is also inverted to a low by Q3C1, applied to the AUTO/MANUAL switch S372, and is used for automatic audio switching when S372 is set to the AUTO position.

The user may select either STEREO or SAP with switch S374. When STEREO is selected, 12 V is applied to the base of Q803 through S374 and R811. The high at the base of Q803 maintains a low at pin 3 of IC801, holding IC801 in the stereo or L−R position (pin 4 connected to 7). Even if an SAP signal is present, the low from Q3C1 (indicating a SAP broadcast) is overridden by the 12 V from S374, and only the L−R stereo signal passes to the dbx circuits, as described in Sec. 3−2.2.

If the user sets S374 to SAP and S372 is in AUTO, Q803 is held on by the high at the collector of Q3C1. With Q803 on, the collector of Q803 and pin 3 of IC801 go low, and stereo is selected (pins 4 and 7 of IC801 are connected). However, if SAP is being broadcast, the collector of Q3C1 goes low. This turns Q803 off and applies a high to pin 3 of IC801, thus selecting SAP (pins 4 and 2 of IC801 connected). If the SAP broadcast ends, IC801 is switched automatically back to stereo.

When the AUTO/MANUAL switch S372 is set to the MANUAL position, Q803 is controlled solely by S374. With S374 set to STEREO, Q803 is turned on, pin 3 of IC801 goes low, and stereo is selected. With S374 set to SAP, Q803 is turned off, pin 3 of IC801 is high, and SAP is selected.

Note that if S372 is set to MANUAL, with S374 set to SAP, and there is no SAP signal being broadcast, there will be no audio applied to the amplifiers and loudspeakers. This is because pin 3 of IC801 remains high (Q803 not turned on) and keeps pin 4 of IC801 connected to pin 2 (where no SAP signal is being broadcast).

Also note that the logic from S374 and S372 is applied to the base of Q3D0 and the audio amplifier input switching circuits, which we describe next.

3−2.5 *Audio amplifier input switching*

Figure 3−6 shows the audio amplifier input switching circuits in simplified form, together with a truth table for the switching functions. The selection of the desired audio signal is determined by two four-position switches within IC351. Since only three audio sources are available (stereo, SAP, and external), only three of the four positions are actually used in this application.

The positions of the switches within IC351 are controlled by the logic at pins 9 and 10 of IC351. In turn, the logic is determined by the settings of S372/S374 and the presence or absence of stereo/SAP signals. The truth table on Fig. 3−6 shows the logic for each combination of S372/S374 and signal conditions. Note that the EXTERNAL signal is not part of the stereo decoder or SAP circuits.

As an example, assume that S372 is set to AUTO, S374 is set to SAP (as shown in Fig. 3−6), and there is a SAP broadcast. Under these conditions, Q3D0 is not turned on, and both pins 9 and 10 of IC351 are set to high by the 12 V at the junction of R3M4/R3M6. This connects pin 11 of IC351 to pin 13, and pin 4 of IC351 to pin 3, to pass the SAP signal.

Mode		Broadcast Signal	Input Logic		Audio Amplifier Input
S372			Pin 9	Pin 10	
	S374				
AUTO		Stereo	H	L	Stereo
	STEREO	SAP	H	L	Stereo
AUTO		Stereo	H	L	Stereo
	SAP	SAP	H	H	SAP
MANUAL		Stereo	H	L	Stereo
	STEREO	SAP	H	L	Stereo
MANUAL		Stereo	H	H	No sound
	SAP	SAP	H	H	SAP
		EXTERNAL	L	H	External input

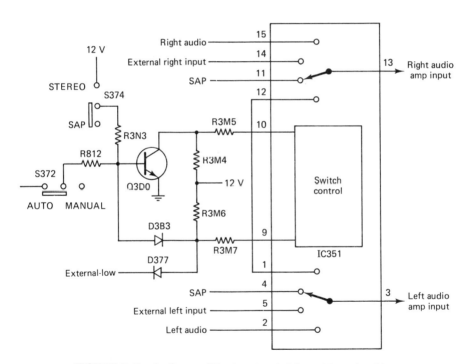

FIGURE 3-6   Audio amplifier input switching with truth table.

If the SAP broadcast ends, Q3D0 turns on, the junctions of R3M4/R3M5 is shorted to ground, and pin 10 of IC351 goes low. This connects pin 15 of IC351 to pin 13, and pin 2 of IC351 to pin 3, to pass the L – R stereo and/or mono signals.

3-3. TYPICAL TEST/ADJUSTMENT PROCEDURES

All of the notes described in Sec. 1-7.6 apply to the following test/adjustment procedures. There are four basic adjustments required for the circuits described in Sec. 3-2. These include adjusting the L−R decoder and SAP detector PLL circuits, so that the corresponding VCO is locked to the incoming signals, and adjusting the level of the stereo and SAP signals to get proper separation and/or balance when compared to the mono L+R signal.

Keep in mind that there may be additional controls in some Mitsubishi circuits (such as those described in Chapter 4), and some of the controls described here do not exist in other circuits. Similarly, there are many approaches for adjustment of these controls. The following paragraphs describe some typical approaches for adjustment of the controls discussed in Sec. 3-2.

3-3.1   L−R decoder adjustment

Figure 3-7 is the adjustment diagram. The purpose of the L−R decoder IC3A1 is to reinsert a carrier into the AM stereo L−R sidebands at pin 2 and produce corresponding audio output at pin 5. The missing L−R carrier is at 31.468 kHz and is produced by a VCO within the L−R decoder. Since there is no L−R carrier at the decoder input, the VCO is usually locked to the 15.734-kHz pilot (or a multiple).

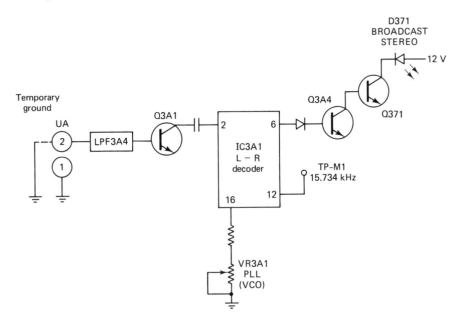

FIGURE 3-7   L−R decoder adjustment diagram.

In the circuit of Fig. 3–7, the VCO can be set by adjustment of PLL adjust VR3A1 connected to pin 16 of IC3A1, and can be monitored at TP-M1 (pin 12). The VCO signal at TP-M1 is 15.734 kHz, even though the VCO operates at a different frequency (typically, 31.734 kHz). Proceed as follows to adjust the L−R decoder.

1. Ground the input to the stereo decoder circuits at pin 2 of the UA connector. This removes all signals to the L−R decoder input, including the 15.734-kHz pilot. If the pilot is present, it is possible that the VCO will lock on to the incoming signal, even though the VCO is not exactly on frequency. Check that the BROADCAST STEREO indicator D371 is off.
2. Connect a frequency counter to TP-M1.
3. Adjust VR3A1 for a reading of 15.734 kHz at TP-M1.
4. Remove the ground from pin 2 of the UA connector. Check that the TP-M1 reading remains at 15.734 kHz.

3–2.2 L−R separation adjustment

Figure 3–8 is the adjustment diagram. The purpose of this adjustment is to set the level of the stereo L−R signal in relation to the mono L+R signal. Both signals are combined in the Q3A6 matrix. If the L−R signal is low in relation to L+R, you will hear only mono. If the L−R is high in relation to L+R, you will hear both signals but there will be poor separation between left and right audio (the audio will sound like mono even though stereo is present).
Proceed as follows to adjust the L−R separation.

1. Apply a modulated L−R signal to pin 2 of the UA connector. The BROAD-CAST STEREO indicator D371 should turn on. Use the modulated signal from the MTS generator described in Chapter 2. Use a modulation frequency of 300 Hz unless otherwise specified in the service literature. Refer to Sec. 2–6.9 for further information on the separation adjustment.
2. Monitor TP-M2 and TP-M3 for the 300-Hz signal, using an audio voltmeter.
3. Adjust VR3A2 for the *correct voltage level* at TP-M2 and TP-M3. Always use the values specified in the service literature (for both input amplitude and output at the test points). Keep in mind that this adjustment can be *critical* in producing good stereo sound.
4. If you cannot find a separation adjustment procedure in the service literature, use the following as an *emergency procedure only.*
5. Set VR3A2 to the full counterclockwise position, so that L−R is zero. Apply an L+R signal with 300-Hz modulation at pin 2 of the UA connector. Note the voltmeter reading at TP-M2 and TP-M3. This is the mono L+R signal.
6. Remove the L+R signal and apply L−R with 300-Hz modulation at pin 2 of the UA connector.

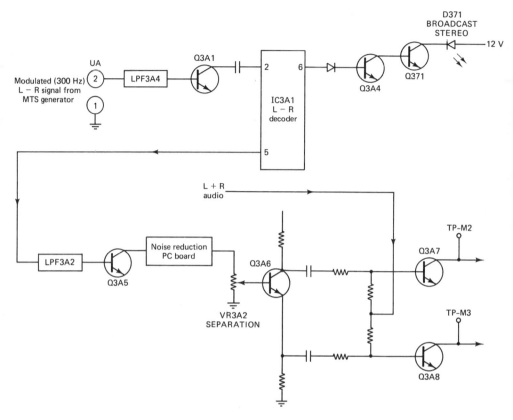

FIGURE 3–8   L–R separation adjustment diagram.

7. Adjust VR3A2 until the readings at TP-M2 and TP-M3 are the same as with L+R (or just below L+R) in step 5.

8. For a final test, measure stereo separation as described in Sec. 2–6.7. (In fact, some service technicians recommend adjusting VR3A2 to get a given separation, or maximum separation, at the loudspeakers, rather than for a given reading in the decoder circuits.) A stereo separation of 60 dB, or better, is possible on some sets.

3–3.3 SAP detector adjustment

Figure 3–9 is the adjustment diagram. The purpose of the SAP signal detector IC3A3 is to produce a low at output pin 7 when the 78.67-kHz SAP carrier is present at input pin 6. (Output pin 7 remains high when the SAP carrier is not present.) This is done by comparing the incoming SAP carrier with the IC3A3 decoder VCO. When both signals are present and locked in frequency/phase, pin 7 goes low.

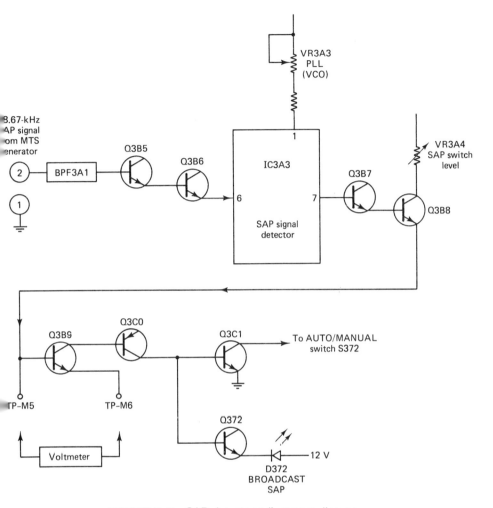

FIGURE 3–9   SAP detector adjustment diagram.

In the circuit of Fig. 3–9, the VCO can be set by adjustment of PLL adjust VR3A3 connected to pin 1 of IC3A3.

Proceed as follows to adjust the SAP detector adjustment.

1. Apply a 78.67-kHz signal to pin 2 of the UA connector. Use the SAP carrier signal from the MTS generator described in Chapter 2.
2. Monitor the voltage across TP-M5 and TP-M6 as shown in Fig. 3–9.
3. Set VR3A3 to full counterclockwise. The BROADCAST SAP indicator D372 should be off. The voltmeter should read about −1.0 V.

4. Adjust VR3A3 clockwise until the voltage at TP-M5 and TP-M6 changes to about +1.4 V. This indicates that pin 7 of IC3A3 is switched to low. You could measure at pin 7 of IC3A3, but the change in signal level is more difficult to detect. Also, by checking at TP-M5 and TP-M6, you confirm operation of Q3B7, Q3B8, Q3B9, and Q3C0 simultaneously.

5. With the SAP carrier still applied at the circuit input (pin 2 of the UA connector), check that the BROADCAST SAP indicator D372 has turned on when the TP-M5/TP-M6 voltage indication changes from −1.0 V to +1.4 V. This confirms operation of both D372 and Q372.

Note that in some Mitsubishi circuits, the SAP detector signal *output level* must also be set, using a control in the emitter of Q3B8. (Refer to Sec. 3–5.2.) However, this is not the case with the circuits described in Sec. 3–2. Also note that the SAP detector signal is sometimes called the SAP *switch signal,* in contrast to the SAP audio signal.

3–3.4   SAP level adjustment

Figure 3–10 is the adjustment diagram. The purpose of this adjustment is to set the level of the SAP signal in relation to the stereo L−R signal. (Not all Mitsubishi decoder circuits have adjustments.) If the SAP and L−R signals are not approximately equal at this stage in the audio path, the user audio output or volume control must be reset each time when switching between stereo and SAP.

If you cannot find an SAP level adjustment procedure in the service literature, use the following as an *emergency procedure only.*

1. Apply an L−R signal with 300-Hz modulation to the input of the decoder circuits (pin 2 of the UA connector).

2. Measure the L−R audio signal at pin 7 of IC801.

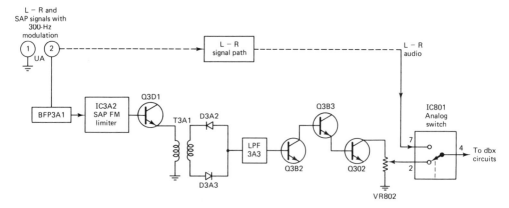

FIGURE 3-10   SAP level adjustment diagram.

3. Remove the L−R signal and apply a SAP signal with 300-Hz modulation to the input of the decoder circuits.

4. Adjust VR802 until the signal at pin 2 of IC801 is the same as the L−R signal at pin 7 of IC801, as measured in step 2.

5. For a final test, measure the level of the signal at the loudspeakers, first with a SAP signal at the decoder input, then with L−R at the input. Use 300-Hz modulation with both signals. The loudspeaker voltages should be approximately equal with L−R and SAP.

Note that in some Mitsubishi circuits, it is not possible to adjust the SAP audio level.

3-4. TROUBLESHOOTING APPROACH

All of the notes described in Secs. 1-7 and 1-8 apply to the following procedures. In each of the following trouble symptoms, it is assumed that an active TV channel has been tuned in, that both stereo and SAP signals are (supposedly) being broadcast (in addition to a mono broadcast), and that the video is good. If you do not have a good picture, the problem is probably in sections ahead of the MTS circuits (such as in the RF or IF circuits).

As a first step, check if any audio is available at the loudspeakers (indicating a mono broadcast). Then check if the BROADCAST STEREO and BROADCAST SAP indicators are turned on (indicating stereo and SAP broadcast signals).

3-4.1 No audio available at the loudspeakers

If there is no audio available at the loudspeakers, but there is a good picture and either the BROADCAST STEREO or BROADCAST SAP indicator is on, the problem is probably in the mono L + R audio path or in the audio amplifiers that follow the MTS circuits. The fact that either indicator is on makes it likely (but not absolutely certain) that the sections ahead of the MTS circuits are good.

Before you plunge into any circuits, make certain that the STEREO/SAP and AUTO/MANUAL switches are in the correct positions and that the corresponding indicators are on. As discussed, if you are in MANUAL and SAP and stereo is broadcast, you will get no sound.

There are several approaches that can be used at this point. Generally, the simplest is to inject a composite signal (with L + R) at the input of the MTS circuits. As shown in Fig. 3-3, this input is at pin 2 of the UA connector.

If the modulated signal is not heard on the loudspeakers with an L + R signal applied, trace through the audio path using signal injection or signal tracing, whichever is most convenient. The L + R path (up to the amplifiers) includes LPF3A4, Q3A1, LPF3A1, Q3A3, Q806, IC804, Q3A7, and Q3A8 shown in Fig. 3-3, and IC351/Q3D0 shown in Fig. 3-6.

If the modulated signal is heard on the loudspeakers, it is reasonable to assume that the audio path through the MTS circuits (Figs. 3–3 and 3–6) and the audio amplifiers is good. The problem is probably ahead of the MTS circuits.

We do not cover the RF and IF sections in this book since we are concerned with MTS or stereo TV. Possible exceptions to this are the sound IF (SIF) circuits between the IF and MTS circuits. Figure 3–11 shows some typical SIF circuits in simplified form.

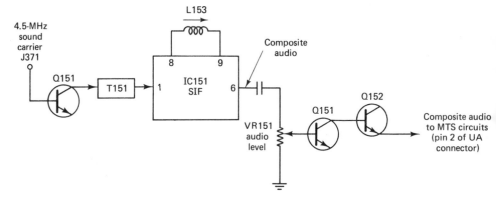

FIGURE 3–11   Typical SIF circuits.

The SIF circuits extract audio from the 4.5-MHz audio carrier and are similar to those found in nonstereo TV, so we will not dwell on them. However, it should be noted that there are two adjustment controls that can affect MTS operation. One control is L153, which sets the frequency of a quadrature detector within IC151. The other control is VR151, which sets the level of audio taken from IC151. The MTS generator described in Chapter 2 can be used to make these adjustments. Use the adjustment procedures found in service literature. Generally, the procedures are the same as for nonstereo TV. However, it may be necessary to set L151 for a given level (such as for the separation adjustment described in Sec. 2–6.9).

To localize trouble within the circuit of Fig. 3–11, inject a 4.5-MHz sound carrier with L+R at J371. If the sound passes when injected at the input of the MTS circuit (pin 2 of UA), but not when injected at J371, the problem is in the SIF circuits of Fig. 3–11. Trace the 4.5-MHz audio carrier to IC151, and audio from IC151 to the MTS input, in the usual manner.

3–4.2 No stereo operation, mono operation is good

If there is mono audio available at the loudspeakers, and the picture is good but there is no stereo operation, check that the STEREO/SAP switch S374 is in STEREO, that the STEREO indicator D373 is on (indicating that stereo has been selected), and that the BROADCAST STEREO indicator D371 is on (indicating that stereo is being broadcast).

If the BROADCAST STEREO is on, the problem is probably in the L−R audio path shown in Figs. 3-3 and 3-4. If the BROADCAST STEREO is off, the problem may be that no stereo is being broadcast! The quickest approach at this point is to inject a composite signal (with L−R) at the input of the MTS circuits (pin 2 of UA).

If BROADCAST STEREO turns on and you get stereo at the loudspeakers, the problem is probably ahead of the MTS circuits. Proceed as described in Sec. 3-4.1. If BROADCAST STEREO does not turn on and there is no stereo at the loudspeakers, the problem is in the L−R audio path. (The same is true if BROADCAST STEREO is on but there is no stereo at the loudspeakers.)

There are several points to consider when tracing through the L−R path, whether signal injection or signal tracing is used. First try correcting the problem by adjustment of the L−R decoder PLL, as described in Sec. 3-3.1. Then try the L−R separation (or level) adjustment of Sec. 3-3.2. These adjustments could cure the problem. If not, simply making the adjustments might lead to the defect.

For example, if there is no 15.734-kHz signal at TP-M1, no matter how VR3A1 is adjusted, IC3A1 is suspect. Similarly, if pin 6 of IC3A1 does not go low when a pilot is injected at the MTS input, IC3A1 is suspect.

Another point to check in the L−R path is at pin 3 of IC801 (Fig. 3-5). If pin is not low, with STEREO/SAP switch S374 set to STEREO, S374, R811, and Q803 are suspect. If pin 3 is low, check at pins 4 and 7 of IC801. The signal should be substantially the same at pins 4 and 7 when pin 3 is low. If not, IC801 is suspect.

Note that pin 3 of IC801 should also go low with S374 set to SAP if there is no SAP signal. If not, the SAP circuits are suspect. Proceed as described in Sec. 3-4.3.

Check at pins 5 and 15 of IC802 (Fig. 3-3). If there is no signal at pin 15, IC802 is suspect. If there is a signal at pin 15 but not at pin 5, IC803 is suspect. Note that the signal at pin 5 should be an amplified version of the signal at pin 15. If not, suspect IC802.

3–4.3 No SAP operation, mono and stereo operation are good

If there is mono and stereo audio, and the picture is good but there is no SAP operation, check that the STEREO/SAP switch S374 is in SAP, that the SAP indicator D374 is on (indicating that SAP has been selected), and that the BROADCAST SAP indicator D372 is on (indicating that SAP is being broadcast).

If the BROADCAST SAP is off, the problem may be that no SAP is being broadcast! If the BROADCAST SAP is on, the problem is probably in the SAP audio path shown in Fig. 3-5.

Note that there are actually two SAP paths. One path is for the SAP audio from pin 6 of IC3A2 to pin 2 of IC801. The other path is for the SAP switch signal from Q3B5 to pin 3 of IC801 and to BROADCAST SAP indicator D372. The quickest approach at this point is to inject a composite signal (with SAP) at the input of the MTS circuits (pin 2 of UA).

If BROADCAST SAP does not turn on and there is no SAP at the loudspeakers, the problem is in the SAP audio paths. (The same is true if BROADCAST SAP is on but there is no SAP at the loudspeakers.)

There are several points to consider when tracing through the SAP paths, whether signal injection or signal tracing is used.

First try correcting problems in the SAP switch-signal path by adjustment of the SAP detector, as described in Sec. 3-3.3. Then try the SAP level adjustment of Sec. 3-3.4. These adjustments could cure the problem. If not, simply making the adjustments might lead to the defect.

For example, if pin 7 of IC3A3 does not go low when there is a SAP signal applied at the input, no matter how VR3A3 is adjusted, IC3A3 is suspect. (IC3A3 is also suspect if pin 7 remains low when the SAP input is removed.)

If pin 7 of IC3A3 does go low when SAP is applied, check that the high appears on the base of Q372. If so, and the BROADCAST SAP indicator D372 does not turn on, D372, R380, and Q372 are suspect. If the base of Q372 does not go high when there is a SAP input (and pin 7 of IC3A3 is low), suspect Q3B7, Q3B8, Q3B9, and Q3C0, as well as R3L0 and R376/R377.

Another point to check in the SAP switch signal path is at pin 3 of IC801. If pin 2 is not high with STEREO/SAP switch S374 set to SAP and BROADCAST SAP indicator D372 on, Q3C1, S372, and Q803 are suspect.

If pin 3 is high, check at pins 2 and 4 of IC801. The signal should be substantially the same at pins 2 and 4 when pin 3 is high. If not, IC801 is suspect.

Note that if AUTO/MANUAL switch S372 is set to MANUAL, the SAP switch signal has no effect on pin 3 of IC810 or on IC351/Q3D0 (shown in Fig. 3-6).

3-4.4 Audio input-switching problems

Problems in audio input-switching circuits (Fig. 3-6) can produce trouble symptoms that appear to be in other circuits. For example, if there is no mono or stereo, it is possible that IC351 has not received a command that selects pins 2 and 15. This could be the fault of Q3D0 or the associated circuit. Similarly, IC351 may receive the correct command but not respond properly. This is the fault of IC351.

Before you condemn the MTS circuits ahead of the audio input switching, check all signals to and from IC351. Compare the signals at pins 9 and 10 of IC351 with the logic shown on the truth table of Fig. 3-6. Then see if the correct input pins are selected.

As an example, set S372 to MANUAL (so that the SAP signal will have no effect) and set S374 to STEREO. Pin 9 of IC351 should go high (through S374, D3B3, and R3M7). Pin 10 should go low (through R3M5 when Q3D0 is turned on). Under these conditions, pins 2 and 15 of IC351 should be selected.

Now set S374 to SAP. Both pins 9 and 10 should go high (through R3M4, R3M5, R3M6, and R3M7). Under these conditions, pins 4 and 11 of IC351 should be selected.

3–5. SOME ADDITIONAL CIRCUITS

The following paragraphs describe some additional MTS circuits found in Mitsubishi television. These circuits are similar to those described in this chapter thus far, but with subtle differences that must be considered in troubleshooting.

3–5.1 L+R and L−R circuits

Figures 3–12 through 3–15 show some additional mono/stereo circuits in simplified form. Compare these circuits with those shown in Fig. 3–3. In the Fig. 3–12 circuits, the L+R signal passes through LPF3A4, Q3A1, LPF3A1, Q3A3, and circuits on a noise-reduction PC board. (The noise-reduction PC board components are not replaceable.) After passing the noise-reduction components, the L+R signal is applied to the Q3A6 stereo matrix.

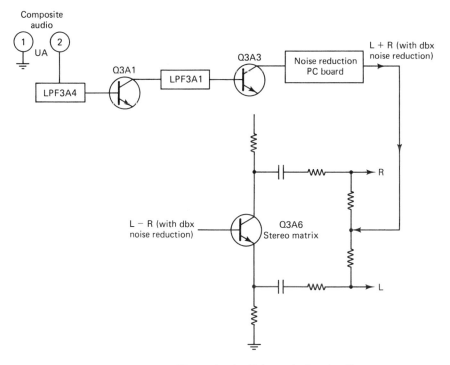

FIGURE 3–12 Alternative L+R (mono) signal path.

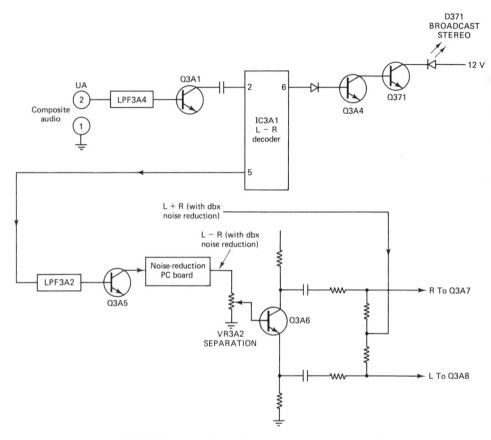

FIGURE 3-13   Alternative L−R (stereo) signal path.

In the Fig. 3-13 circuits, the L−R signal passes through LPF3A4, Q3A1, IC3A1, LPF3A2, Q3A5, noise-reduction circuits, and Q3A6, where L−R signals are combined with L+R in the matrix.

The L and R signals following the matrix (Fig. 3-14) are amplified by Q3A7/Q3A8. Note that Q3A9 is connected across the collectors of Q3A7/Q3A8 through capacitors C3C7/C3D2. Q3A9 is controlled by a front-panel BLEND switch. When BLEND is selected by the user, the front-panel BLEND indicator turns on and the blend function is selected.

Occasionally, broadcast stereo transmissions may contain excessive background noise (hiss), particularly in fringe areas located far from the broadcast station. The dbx noise-reduction system is supposed to eliminate (or minimize) such hiss, but no system can overcome the problems of very weak stereo signals. The blend circuit provides additional noise reduction.

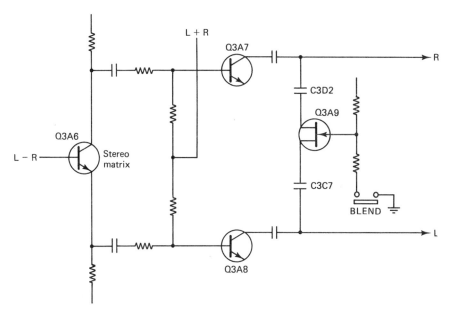

FIGURE 3-14 Typical blend circuits for additional noise reduction.

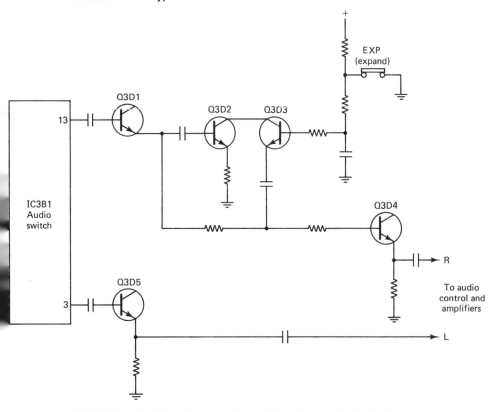

FIGURE 3-15 Typical expand (simulated stereo effect) circuits.

From a troubleshooting standpoint, remember that *blend reduces the amount of separation* between the right and left channels of a stereo broadcast. This is particularly true at the higher (treble) audio frequencies. Blend should be operated only when background noise is objectionable. For normal use, the BLEND switch should be set to the OFF position.

The L and R signals following audio switch IC3B1 (Fig. 3–15) are amplified by Q3D1, Q3D4, and Q3D5. Note that Q3D2 and Q3D3 are controlled by a front-panel EXP (expand) switch. When expand is selected by the user, the front-panel EXP indicator turns on and the expand function is selected.

The audio expand function provides a *simulated stereo effect* when used with monaural broadcasts or with playback from monaural external units (VCRs, videodisc players, etc.). Such external units can be connected to the rear-panel audio inputs.

From a troubleshooting standpoint, remember that audio expand should be used only with mono (or possibly with SAP) but not with stereo. If expand is used with stereo, the audio output will not be true stereo.

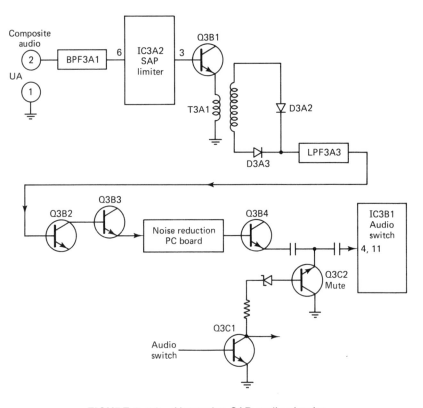

FIGURE 3–16   Alternative SAP audio circuits.

3-5.2 SAP circuits

Figures 3-16 and 3-17 show some additional SAP circuits in simplified form. Compare these circuits with those shown in Fig. 3-5. There are three major differences in the SAP audio circuits of Fig. 3-16. First, the SAP audio is passed through separate SAP circuits on the noise-reduction PC board. Second, the level of the SAP audio is not adjustable (and there is no control comparable to VR802 in Fig. 3-5). Third, the SAP audio is not passed through an L−R/SAP switch (comparable to IC801 in Fig. 3-5). Instead, the SAP audio is controlled by a mute transistor Q3C2.

When Q3C2 is on, the SAP audio is bypassed to ground. Q3C2 is controlled by signals from the SAP switch circuit.

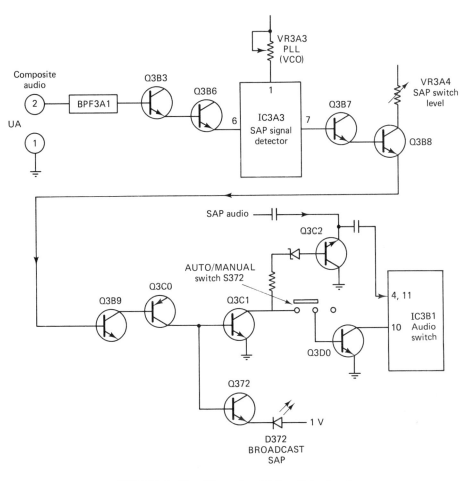

FIGURE 3-17 Alternative SAP switch circuits.

The one major difference in the SAP switch circuits of Fig. 3-17 when compared to those of Fig. 3-5 is that the level of the SAP switch signal can be adjusted by VR3A4 in the circuits of Fig. 3-17. VR3A4 thus sets the level at which the SAP switch signal controls the mute transistor Q3C2. (One advantage of this adjustment is that you can screen out weak SAP signals, if desired.)

When troubleshooting and/or adjusting the SAP switch circuits, make certain to adjust VR3A4 before setting the SAP detector PLL adjustment of VR3A3. The setting of VR3A4 serves as a reference point for adjustment of VR3A3.

4

MITSUBISHI HI-FI/STEREO VCR CIRCUITS

This chapter is devoted to the decoder circuits (again referred to as the MCS circuits) used in typical Mitsubishi VHS hi-fi/stereo VCRs. Such VCRs are capable of recording and playing back both main-program stereo-TV and SAP broadcasts, as well as monaural broadcasts.

Note that hi-fi VCRs differ from conventional stereo or mono VCRs. With hi-fi VCRs, the audio is recorded through heads located on the rotating drum or scanner. This is the same drum/scanner used to rotate the video heads. In Beta, the same heads are used for both audio and video. In VHS, such as that used by Mitsubishi, separate sets of heads are used for audio and video.

With either Beta or VHS, sound is also recorded as a monaural signal with the conventional fixed audio head. This assures complete compatibility with mono VCRs.

In the normal operating sequence, the main-program L−R signal is recorded and played back through the rotating heads (as hi-fi stereo). Simultaneously, the L+R signal, or the SAP signal, is recorded on the conventional head.

The appropriate recording system (stereo L−R or mono L+R) is selected automatically, depending on the type of broadcast signal. However, the choice between main-program L+R or SAP must be selected manually by the user.

4-1. OPERATING (USER) CONTROLS AND INDICATORS

Three front-panel operating (or user) controls and two front-panel indicators are associated with the decoder circuits. Figure 4–1 shows a typical arrangement for the controls and indicators, as well as truth tables.

The STEREO broadcast indicator is turned on only when a stereo broadcast is received. The SAP indicator is turned on only when an SAP broadcast is received.

If the broadcast is in stereo, the left channel of the signal is recorded as channel 1 of the rotating hi-fi sound track, while the right channel is recorded as channel 2. If the broadcast is in mono, both the left and right channels of the signal are recorded on the rotating hi-fi sound track, but as a mono signal. The INPUT SELECT switch must be in the TV position (also called the TUNER position in some cases) for broadcast audio (either mono or stereo) to be recorded on the rotating hi-fi sound track.

If the NORMAL AUDIO switch is set to L+R, main-program material is recorded on the stationary head. If the NORMAL AUDIO switch is set to SAP and there is an SAP broadcast (as indicated by the SAP indicator turning on), SAP material is recorded on the stationary head. If the NORMAL AUDIO switch is set to SAP and there is no SAP material being broadcast (SAP indicator off), the main-program L+R is recorded automatically on the stationary head.

Playback from the rotating heads depends on the position of the AUDIO MONITOR switch and the material recorded on the heads. Playback from the fixed head is always monaural (L+R or SAP).

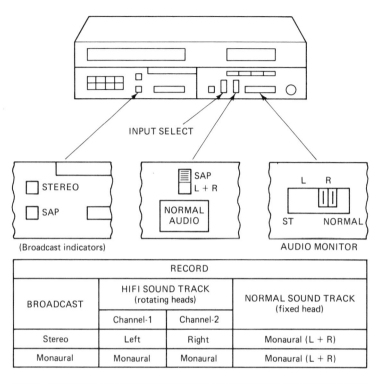

(Broadcast indicators)

AUDIO MONITOR

RECORD			
BROADCAST	HIFI SOUND TRACK (rotating heads)		NORMAL SOUND TRACK (fixed head)
	Channel-1	Channel-2	
Stereo	Left	Right	Monaural (L + R)
Monaural	Monaural	Monaural	Monaural (L + R)

PLAYBACK			
AUDIO MONITOR switch	Material recorded on rotating heads		L + R or SAP (fixed head)
	Two audio channel TV	Monaural TV	
ST	Stereo	Monaural	Monaural
L	Left	Left	Monaural
R	Right	Right	Monaural
NORMAL	Monaural	Monaural	Monaural

FIGURE 4-1 Typical arrangement for operating controls and indicators.

73

4–2. CIRCUIT DESCRIPTIONS

All of the notes described in Sec. 1–7 apply to the following descriptions.

4–2.1 Basic hi-fi recording system and audio signal conversion

Before we get into the details of the decoder circuits used in hi-fi VCRs, let us review the basics of hi-fi recording and audio signal conversion found in VCRs. It is assumed that you are already familiar with the principles of VCRs. If not, read the author's best-selling *Complete Guide to Videocassette Recorder Operation and Servicing* (Englewood Cliffs, N.J.: Prentice-Hall, Inc., 1983) and *Complete Guide to Modern VCR Troubleshooting and Repair* (Englewood Cliffs, N.J.: Prentice-Hall, Inc., 1985).

High-fidelity recording is made possible by a combination of rotating audio heads (effectively increasing the tape speed), and a special recording process described as *depth-layer* or *multiple-layer* recording. Figure 4–2 shows the basic principles involved.

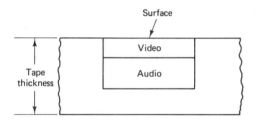

FIGURE 4–2   Depth-layer or multiple-layer recording.

The techniques shown in Fig. 4–2 permit both video and audio information to be deposited on a single recorded track. Initially, the audio is recorded deep into the tape's magnetic medium. The video signal is then recorded on the surface of the tape. In effect, the video is superimposed above the audio signal.

As is the case with VCR video, an azimuth recording technique is used for hi-fi audio. (Both the audio and video signals on a given track are recorded at opposing azimuths.) With azimuth recording, any interaction between the two signals during playback is at a minimum.

Figure 4–3 shows the relative positions of the video and audio heads mounted on the rotating drum of a VHS VCR. As dictated by the VHS format, the two video heads are spaced 180° apart, and each audio head is positioned 90° from the corresponding video head.

As shown in Fig. 4–4, the video and audio tracks could not be superimposed without some form of compensation, due to the movement of the tape and the 90° differential between the video and audio heads. If this condition were allowed

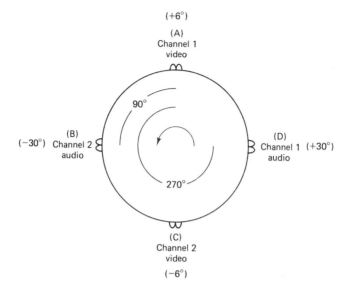

FIGURE 4-3   Relative positions of the video and audio heads on a rotating drum of a hi-fi VHS VCR.

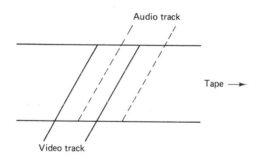

FIGURE 4-4   Track displacement with normal tape travel.

to exist, the succeeding audio track would erase most of the previously recorded video track.

To prevent this condition, the video and audio head heights are displaced, as shown in Fig. 4-5. Each audio head (which records first) is mounted on the drum 29 μm higher than the corresponding video head. Such a displacement assures that the audio and video tracks align properly, one atop the other.

The table in Fig. 4-5 specifies the azimuth and recording track width for each of the individual recording heads. As shown, the video-head azimuths remain identical to those of conventional VHS recorders ($+6°$ and $-6°$). The azimuth of the audio heads are expanded considerably ($+30°$ and $-30°$).

Head	Track Width	Azimuth
Channel 1 video	30 μm	+6°
Channel 2 video	30 μm	−6°
Channel 1 audio	24 μm	+30°
Channel 2 audio	24 μm	−30°

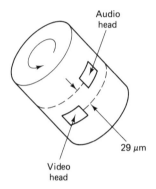

Audio head

29 μm

Video head

FIGURE 4-5   Relationship of audio and video head height.

Figure 4–6 shows the hi-fi recording technique used in a 2H speed mode. The channel 1 video track is recorded atop the channel 2 audio track. Since the VHS format allows a 58-μm video track width in the 2H mode, and the widest track produced by a given head is equal to only 30 μm, guard bands are produced.

Note that in the 2H mode, corresponding video and audio signals are recorded at opposing azimuths (+6° for the video track and −30° for the corresponding

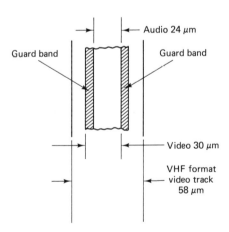

Audio 24 μm

Guard band   Guard band

Video 30 μm

VHF format
video track
58 μm

FIGURE 4-6   Hi-fi recording technique used in a 2H-speed mode.

audio track, and vice versa). This results in a total azimuth difference of 36°. As a result, there is little or no interaction between the two signals.

During the 2H hi-fi recording mode, the video signal recorded atop a given audio track is actually supplied by the adjacent video head. For example, referring to Fig. 4-3, when audio head B records channel 2 audio, video head A records channel 1 video, superimposing the video track onto the audio track. Then audio head D records channel 1 audio, with video head C recording channel 2 video.

Due to the tape movement and the 90° differential between head-mounting positions, the video track to be recorded would appear to fall atop one-half the breadth of the audio track. This is prevented by raising the head 29 μm, as discussed. As a result, the video 1 track is properly superimposed above the channel 2 audio track, and the video 2 track is superimposed directly above the channel 1 audio track.

Figure 4-7 shows the audio signal-conversion spectrum for VHS hi-fi VCRs. As in the case of the video signal, the audio signal to be recorded is converted to an FM signal at a relatively high frequency. Two separate audio signals are developed, one for the right channel and one for the left channel. The two signals are then combined and directed to the audio heads.

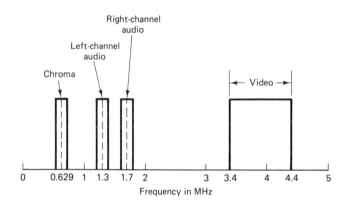

FIGURE 4-7   Audio signal-conversion spectrum for VHS hi-fi VCRs.

The right-channel FM center frequency is 1.7 MHz, and the left-channel frequency is at 1.3 MHz. Nominal deviation for both channels is +50 kHz, with a maximum allowable deviation of +150 kHz. The selection of these audio frequencies places the audio FM signal between the down-converted VHS chroma signal (629 kHz) and the VHS video signal (3.4 to 4.4 MHz), as shown in Fig. 4-7.

In addition to normal FM recording techniques, the signal is processed with dbx noise reduction, similar to that described for other stereo-TV circuits. The noise-reduction feature improves the signal-to-noise ratio of the reproduced sound.

To assure compatibility with existing VHS VCRs and existing tapes recorded for use with such VCRs, a hi-fi VCR also has the conventional fixed audio head. The fixed audio head, sometimes called the *linear head,* records a separate audio track along the upper edge of the video tape.

4-2.2 Basic overall hi-fi VCR sound circuits

Figure 4–8 shows the basic sound circuits of a hi-fi VCR. There are two basic audio paths. One path supplies signals to the conventional stationary audio head (linear head). The second path supplies signals to the circuits used by the two hi-fi heads. Signals to be recorded on the rotating hi-fi heads may be derived from two sources: the external audio inputs or the television broadcast signal.

The TV signal from the 4.5-MHz detector may assume one of two forms: normal TV monaural sound or composite stereo sound. The TV signal may or may not also contain an SAP signal.

The signal from the 4.5-MHz detector is directed to the *stereo decoder circuit.* If stereo is being broadcast, the signal is decoded and the individual right and left audio signals are applied to the *input-select circuit.* If the TV sound transmission is monaural, the stereo decoder produces identical monaural signals at both the right- and left-channel inputs to the input-select circuit.

The input-select circuit selects either TV sound or external audio from the external inputs and directs the selected sound to the *record hi-fi amplifier.* The sound signal is subjected to both fixed and dynamic preemphasis (to improve the *S/N* ratio), as well as amplification, and is applied to the *FM modulator.* A sample of the signal from the record hi-fi amplifier is also directed to the *playback hi-fi*

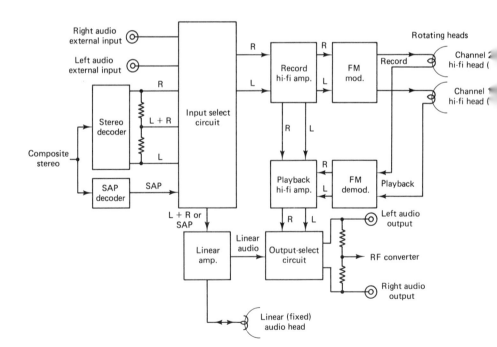

FIGURE 4–8 Basic sound circuits of a hi-fi VCR.

amplifier. The sample signal is then applied to the audio output jacks and the RF converter through the *output-select circuit.*

The right and left audio signals modulate two individual FM carriers in the FM modulator. The resulting FM signals are combined and applied to the hi-fi heads (rotating heads) for recording on tape. Since both audio signals used in hi-fi VCR recording are FM, there is no need for a bias oscillator.

During playback, the hi-fi heads pick up the previously recorded audio. The two audio signals are combined, amplified, frequency separated, and directed to the *FM demodulator circuits.* After demodulation, the right and left audio is applied to the *playback hi-fi amplifier.*

The playback amplifier amplifies both audio signals, as well as performing both fixed and dynamic deemphasis on the audio signals. This returns the audio to the original form, with a hi-fi-quality S/N ratio. The restored audio is directed to the audio output jacks and to the RF converter through the output-select circuit.

There are also two possible signal sources for linear audio head (stationary head). Portions of each of the left and right outputs of the stereo decoder are added together to form a monaural L + R signal for the linear head. If SAP is transmitted, the output of the *SAP decoder* may also be directed to the linear head.

The composite stereo signal applied to the stereo decoder is also applied to the SAP decoder. If an SAP signal is present, the SAP signal is decoded and SAP audio is applied to the input-select circuit. Either mono L + R audio or SAP audio, as selected by the input select circuit, is applied to the linear head through the *linear amplifier.* The linear amplifier is essentially the same as those used in conventional non-hi-fi VCRs.

During playback, the audio signal picked up by the linear audio head (either L + R or SAP) is processed by the linear amplifier and directed to the output-select circuit. If desired, the output-select circuit directs the audio from the linear head to the audio output jacks and the RF converter.

4–2.3 Overall stereo decoder

Figure 4–9 shows the basic stereo and SAP decoder circuits. Note that these circuits are similar but not identical to the stereo-TV/SAP circuits described in Chapter 3.

Individual components of the composite stereo signal may take any of three different paths. Which path a specific component takes is determined by low-pass and bandpass filters (LPFs and BPFs) in the input circuit.

The L + R audio is selected by an LPF designed to pass only those L + R signal frequencies below 15 kHz. The L + R signal (mono audio) is amplified and applied to the audio matrix.

A second LPF passes all frequencies below 46 kHz, which includes the L + R audio, L – R sidebands, and the pilot signal. These signals are all applied to the L – R decoder.

The L + R signal (because of its low frequency) is rejected by the L – R decoder. (Only the L – R sidebands and the pilot signal are used in the L – R decoder.) The

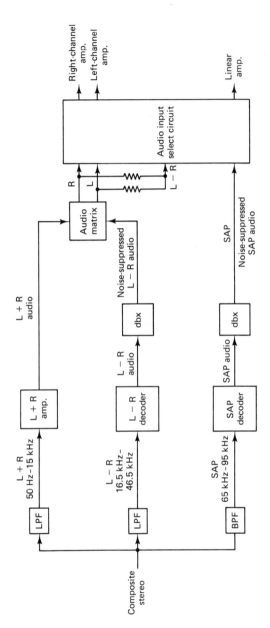

FIGURE 4-9 Basic stereo and SAP decoder circuits.

15.734-kHz pilot signal is used to synchronize an oscillator within the L−R decoder. The oscillator reproduces the suppressed 31.468-kHz L−R subcarrier, thus permitting subsequent decoding of the L−R sidebands.

The output of the L−R decoder is applied to the audio matrix circuit through a dbx noise-reduction circuit. The L−R and L+R signals are combined in the matrix to form individual left and right audio channels. The left- and right-channel audio signals are applied to the audio input-select circuit and are then selected for recording on tape by the rotating hi-fi audio heads.

Samples of each audio channel are added to form the L+R monaural audio signal. This signal is recorded by the rotating hi-fi audio heads if no stereo is broadcast. The L+R signal is also applied to the stationary or linear head when SAP is not broadcast (or selected).

If a SAP signal is present, the SAP audio is passed through the BPF. This BPF passes signals from 65 to 95 kHz and thus rejects the L+R and L−R signals. The SAP signal is decoded by the SAP decoder and is applied to the audio input-select circuit through a separate dbx noise-reduction circuit. The SAP signal can be recorded on the stationary linear head, in place of the L+R, when so selected by the front-panel NORMAL AUDIO control (Sec. 4–1).

4–2.4 L+R circuits

Figure 4–10 shows the L+R circuits in simplified form. The composite stereo signal is amplified by Q3001 and applied through filters to the stereo (L−R), mono (L+R), and SAP signal paths. The L+R and L−R signals are rejected by BPF3Z1 but are passed to Q3020, where both signals are amplified. The L+R signal is separated from the L−R signal at the emitter of Q3020. The L+R signal is applied to amplifier IC3008 through LPF3Z3 and Q3002/Q3003.

The L+R audio from IC3008 is applied to the junction of resistors R3159 and R3160. Both resistors are part of a matrix circuit used to combine the L+R and L−R signals.

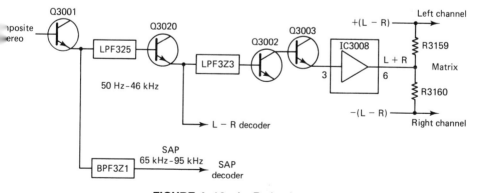

FIGURE 4–10 L+R circuits.

4–2.5 L–R decoder

Figure 4–11 shows the L–R decoder in simplified form. Most of the L–R functions are performed within IC3001. These functions include detecting the presence of a stereo signal (pilot) and decoding the L–R sidebands to produce L–R audio. If the broadcast is in stereo, IC3001 produces a command at pin 6 to turn on the STEREO indicator D521.

The L–R sidebands and the pilot are applied to pin 2 of IC3001 from the collector of Q3020. The L–R sidebands pass to an L–R decoder within IC3001, while the pilot is amplified by a preamp. The amplified pilot signal is applied to two comparators (PCL and PCV) through external C3015.

The VCO within IC3001 operates at four times the pilot frequency, or 62.936 kHz. The VCO output is divided twice by 2, resulting in a 31.468-kHz output (applied to the L–R decoder as a substitute carrier) and a 15.734-kHz output (applied to both PCV and PCL comparators as a second input). When a stereo signal is transmitted, the pilot and the divided-comparison signal from the VCO are compared in both PCV and PCL.

The PCL comparator (phase comparator lamp) generates a signal that is filtered by an LPF, and amplified by a lamp driver, to produce a low at pin 6 of IC3001. If there is no pilot, pin 6 of IC3001 remains high.

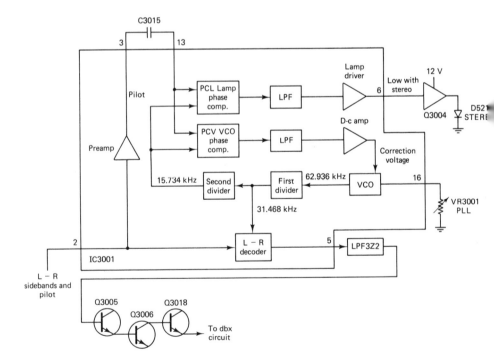

FIGURE 4–11 L–R decoder circuits.

The low at pin 6 of IC3001 turns on transistor Q3004, which, in turn, applies 12 V to the front-panel STEREO indicator D521. As a result, STEREO indicator D521 turns on when a pilot signal is present (indicating a stereo broadcast). If pin 6 is high (no pilot, indicating no stereo), Q3004 and D521 remain off.

The PCV comparator (phase comparator VCO) develops a correction voltage if a frequency or phase error exists between the pilot signal and the VCO comparison signal. The correction voltage is filtered, amplified, and applied to the VCO to correct any frequency or phase error. VR3001, at pin 16 of IC3001, is used to set the free-running frequency of the VCO to four times that of the pilot frequency.

The output of the VCO is applied to a divide-by-2 circuit, reproducing a phase- and frequency-corrected duplicate of the original L−R subcarrier. The resulting 31.468-kHz signal is applied to the L−R decoder, along with the L−R sidebands from pin 2 of IC3001.

The L−R sidebands are decoded, and the resulting L−R audio is available at pin 5 of IC3001. The L−R audio is passed through LPF3Z2, where all unwanted signals above 15 kHz are removed. The L−R audio is then amplified by Q3005/Q3006/Q3018 and applied to the dbx circuits.

4–2.6 dbx noise-reduction format

As discussed in Chapter 1, the dbx noise-reduction format used in stereo TV is the most complex part of the system. However, at the VCR, virtually all of the dbx circuits are contained within one IC. As is the case with any IC, you cannot get inside to test or adjust internal components of the IC. Further, if any component or function within the IC is defective, the entire IC must be replaced. However, there are certain discrete components in the dbx system that can be replaced individually.

Before we get into the actual dbx circuits found in a typical VCR, it is recommended that you review the dbx noise-reduction format described in Chapter 1. The dbx circuits described here are essentially the same as the dbx circuits covered thus far in this book. However, the approach is somewhat different.

4–2.7 dbx circuits

Figure 4–12 shows the dbx circuits in block form. As discussed in Chapter 1, the dbx circuits in the TV set or VCR must sense the amplitude of the L−R signal to determine if the signal requires compression or expansion. The dbx circuit of Fig. 4–12 uses two *rms detectors* to monitor signal amplitude.

A discrete-component BPF passes only those L−R signal frequencies within the *spectral band* (4 to 15 kHz) and applies signals within this band to the *spectral rms detector* (which detects the signal amplitude). Simultaneously, a second discrete-component BPF passes those L−R signal frequencies within the *wideband* (100 Hz to 4 kHz), and applies signals within this band to the *wideband rms detector* (which detects the signal amplitude).

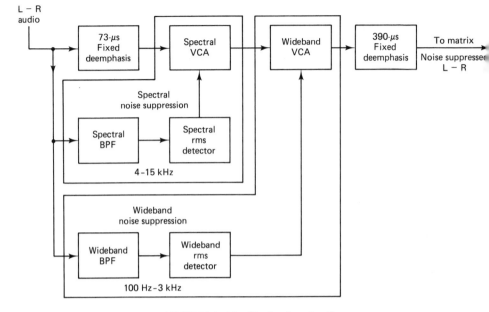

FIGURE 4-12   Basic dbx circuits.

The complete L − R audio signal is deemphasized by a fixed 73-μs deemphasis network. The deemphasized L − R signal is then applied to the *spectral VCA* (voltage-controlled amplifier). (Note that in some dbx ICs, the VCAs are called CCAs, or current-controlled amplifiers).

The gain of the spectral VCA is controlled by the detected level of the spectral rms detector. The signals from 4 to 15 kHz are either compressed or expanded (depending on the respective amplitude) by varying the gain of the spectral VCA. The L − R signal with spectral compression is then directed to the *wideband VCA*.

The gain of the wideband VCA is controlled by the wideband rms detector. The signals from 100 Hz to 3 kHz are compressed by varying the gain to the wideband VCA. The L − R signal from the wideband VCA is deemphasized by a fixed 390-μs deemphasis network, and is applied to the matrix circuit for mixing with the L + R audio signal.

Figure 4–13 shows the fixed 73-μs deemphasis network, as well as the discrete-component BPFs, in simplified form. L − R audio from the emitter of Q3018 is applied to the spectral VCA at pin 18 of IC3004. R3097 and C3136, at the base of Q3042, form the fixed 73-μs deemphasis network.

Q3041 and the associated circuit components form the spectral BPF. L − R signals in the range 4 to 15 kHz are applied to the spectral rms detector at pin 20 of IC3004.

Capacitors C3148/C3149 and resistors R3110/R3111 form the wideband BPF. L − R signals in the range 100 Hz to 3 kHz are applied to the wideband rms detector at pin 3 of IC3001.

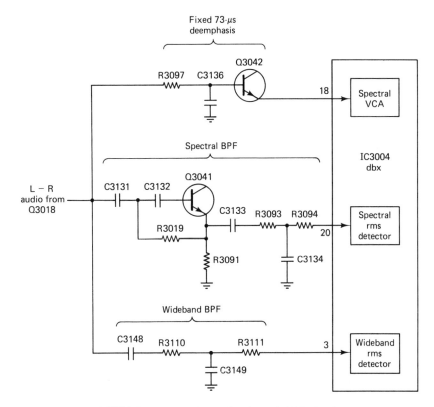

FIGURE 4-13 dbx deemphasis and BPF circuits.

Figure 4-14 shows the dbx circuits within IC3004 in simplified form. The two rms detectors receive power from a constant-current generator within IC3004. The output of the constant-current generator is adjustable by VR3006, the L−R timing control, connected at pin 1 of IC3004. The setting of VR3006 determines the amount of control output from the rms detectors for a given signal amplitude.

Spectral processing is performed by the spectral VCA at pin 18 of IC3004, controlled by the output of the spectral rms detector. The L−R signal is then applied to an L−R amplifier IC3005 through C3143 and an op-amp within IC3004. VR3007 sets the gain of the op-amp, and is sometimes called the *variable deemphasis* or VD control.

The amplified L−R signal is applied to the wideband VCA at pin 5 of IC3004. Wideband processing is performed by the wideband VCA, controlled by the output of the wideband rms detector.

The processed L−R signal is then applied to another op-amp within IC3004 through C3151. The op-amp output at pin 8 of IC3004 is applied to Q3021 in the matrix circuit. C3153 and R3113, between pins 7 and 8 of IC3004, produce the required 390-μs fixed deemphasis.

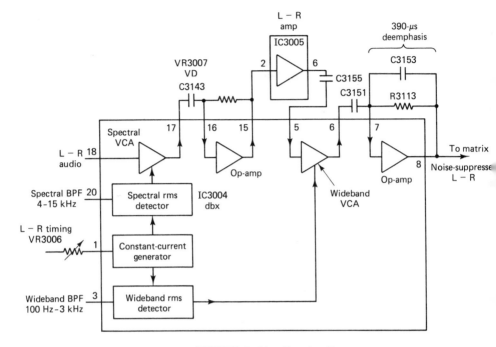

FIGURE 4-14   dbx circuits.

4-2.8   Combined L+R and L-R signal path troubleshooting

Figure 4-15 shows the L + R and L − R audio paths in simplified form. Compare this to the paths shown in Fig. 3-3. The two major differences in the Fig. 4-15 circuits are that the L − R gain can be adjusted, and the dbx circuits are adjustable.

The L − R gain is set by adjustment of VR3002 in the emitter of Q3006, ahead of the dbx circuits. This is not to be confused with the separation adjustment R3010 after the dbx circuits, although both controls set the L − R signal level.

The dbx circuits are set by adjustment of VR3006 and VR3007 connected to dbx IC3004. VR3006 sets the amount of control output from the rms detectors within IC3004 for a given signal amplitude. VR3006 is sometimes called the *L − R timing control.* VR3007 sets the gain of the spectral op-amp within IC3004. VR3007 is sometimes called the *VD, or variable deemphasis control.*

Figure 4-15 is provided here as an aid in troubleshooting. Although we describe troubleshooting in Sec. 4-4, following are some thoughts on using Fig. 4-15.

If there is no stereo, but mono operation is good, the problem is in the L − R path, with IC3001 and IC3004 being the likely suspects. The problem could also be improper adjustments of VR3001, VR3002, VR3006, VR3007, or VR3010. Signal tracing (or signal injection) through the entire L − R path should lead to the problem.

If there is excessive background noise, this is probably caused by problems

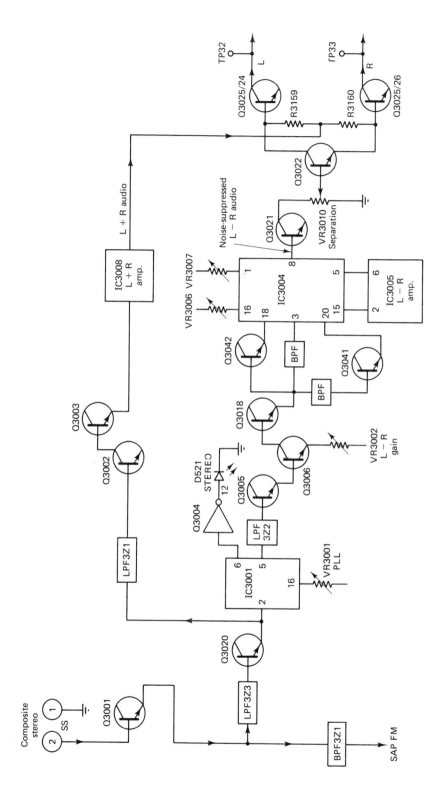

FIGURE 4–15 Combined L+R and L−R signal paths.

in the dbx noise-reduction circuits. IC3004 is the most likely suspect, although the bandpass filters at the inputs to IC3004 are also suspect, since the filters separate the frequencies to be processed by IC3004.

If there is no sound during a monaural broadcast, and distorted sound during a stereo broadcast, this is probably caused by problems in the L + R audio path. During a stereo broadcast, the L − R signal would produce audio. However, if the left- and right-channel audio signals are identical, the L − R signal falls to zero at that instant. As a result, the reproduced sound is erratic and/or distorted, since there is no L + R signal at the matrix.

4–2.9 SAP circuits

Figure 4–16 shows the SAP circuits in block form. The SAP circuits can be divided into two major sections or signal paths.

One section is the *demodulation* signal path that decodes the SAP audio. This audio is applied to the audio input-select circuit through the SAP dbx processing.

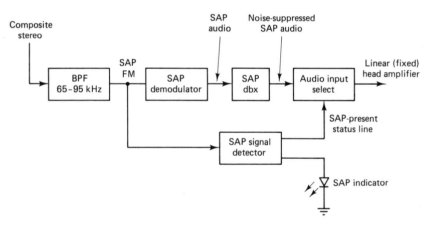

FIGURE 4–16 Basic SAP circuits.

The other section is the *signal-detect* path that turns on the front-panel SAP indicator, and produces an SAP-present signal also applied to the audio input-select circuit. The SAP-present signal permits the user to select SAP for recording by the stationary linear audio head, if desired (instead of L + R).

4–2.10 SAP demodulation signal path

Figure 4–17 shows the SAP demodulation signal path in simplified form. The composite stereo signal is applied to BPF3Z1, which passes only those signals with frequencies between 65 and 95 kHz. The output of BPF3Z1 is applied to the SAP

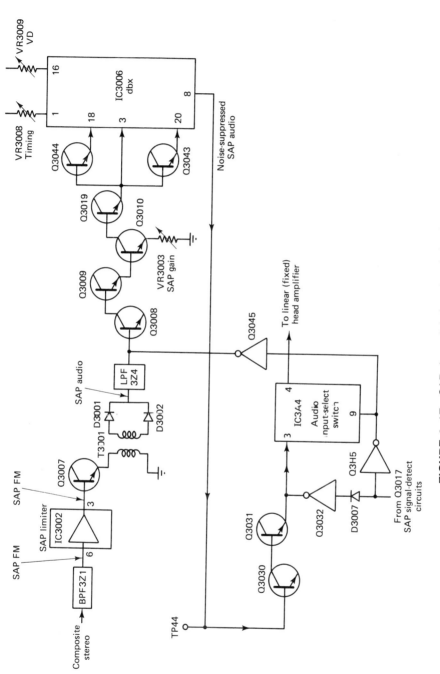

FIGURE 4-17 SAP demodulation signal circuits.

89

signal-detect path (Sec. 4–2.11) and to pin 6 of IC3002, which act as a limiter. The SAP FM from pin 3 of IC3002 is applied to the SAP demodulator (consisting of T3001, D3001, and D3002) through Q3007.

SAP audio from the SAP demodulator is applied to the SAP dbx processing circuits (at Q3044) through LPF3Z4, Q3008, Q3009, Q3010, and Q3019. Note that VR3003 in the emitter of Q3010 sets the level of the SAP audio signal.

Operation of the SAP dbx circuit is essentially the same as that in the L–R circuit (as described in Secs. 4–2.6 and 4–2.7), so we will not repeat the circuit descriptions here.

Although we described troubleshooting in Sec. 4–4, here is a thought on troubleshooting the SAP demodulation signal path. What may appear to be a loss of SAP audio, and a defect in the SAP demodulation path, could be due to a problem in the SAP signal-detect path (Sec. 4–2.11). For example, if the SAP signal-detect circuit does not apply the SAP-present signal to the audio input-select circuit, the SAP signal cannot be directed to the linear audio head (even though the NORMAL AUDIO switch is set to SAP).

4–2.11   SAP signal-detect path

Figure 4–18 shows the SAP signal-detect path in simplified form. The heart of the SAP signal-detect circuit is PLL IC3003. The SAP FM signal is applied at pin 6 of IC3003 through BPF3Z1, Q3011, and Q3012.

The VCO in PLL IC3003 does not oscillate at the SAP carrier frequency of 78.67 kHz. Instead, the VCO oscillates at 44 kHz. If the input to IC3003 is 44 kHz, the output of the phase detector is zero. However, since the SAP signal extends from 46 to 95 kHz, the SAP signal is always higher than the VCO frequency.

As shown by the response curve of Fig. 4–18, the phase detector produces a negative voltage when the incoming signal is higher than the VCO. This negative voltage, generated only when SAP is present, is amplified and filtered. The negative or low output from IC3003 at pin 7 is applied through Q3013, Q3014, Q3015, Q3016 and appears as a high at the collector of Q3016.

The high at Q3016 drives Q3033 into conduction, turning the SAP indicator LED D520 on and informing the user that a SAP signal is being broadcast.

The high at Q3016 is also applied to Q3017, and drives the output of Q3017 low. This output is a status command (called $\overline{SAP}$) applied to Q3H5 in the audio input-select circuit (Fig. 4–17). If the $\overline{SAP}$ line does not go low, the output of the SAP demodulator is not applied to the linear audio head, making it appear that no SAP signal is present. This point should be remembered when troubleshooting for a loss of SAP.

The high at Q3017 is inverted to a low by Q3032, and applied to pin 3 of IC3A4, muting the SAP audio (Fig. 4–17). The Q3017 high is also inverted by Q3045 to a low at the base of Q3008, also muting the SAP audio.

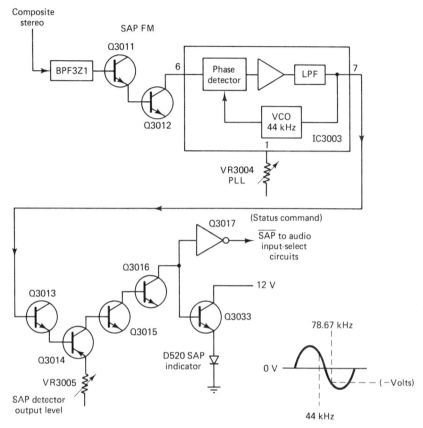

FIGURE 4–18   SAP signal-detect circuits.

4–2.12   Audio input-select circuits

Figure 4–19 shows the audio input-select circuits in simplified form. The audio input-select circuit is responsible for selecting the audio signals to be directed to the rotating hi-fi heads and to the stationary linear head. The user controls involved are the INPUT SELECT switch S3A1 and the NORMAL AUDIO switch S3A3.

The bulk of the audio input-select circuits are within IC3A4. Left-channel audio is output from pin 14 of IC3A4, while right-channel audio is taken from pin 15. The linear head amplifier is supplied from pin 4 of IC3A4 and may be either L + R (mono) or SAP audio.

The right and left outputs of the circuit are controlled by the logic at pins 10 and 11 of IC3A4, taken from S3A1. Both the EXT. AUDIO and EXTERNAL positions of S3A1 select the external audio inputs of the VCR.

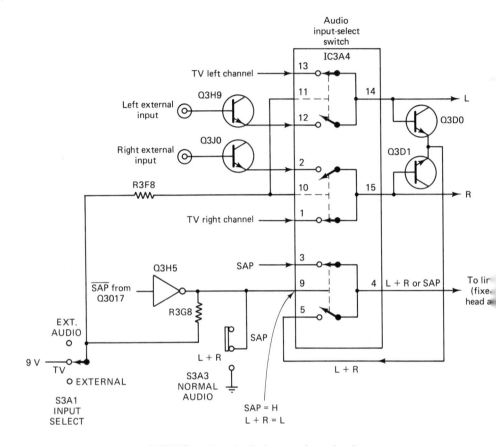

FIGURE 4-19 Audio input-select circuits.

When in the EXT. AUDIO or EXTERNAL position, S3A1 opens the path to the 9-V supply, and the logic at pins 10 and 11 goes low. This causes the switches within IC3A4 to connect pins 2 and 15, as well as pins 12 and 14, and passes the external audio to the hi-fi amplifier inputs.

In the TV position of S3A1, the logic at pins 10 and 11 of IC3A4 goes high (+9 V), opening the analog switches to the external audio input, and closing the switches to the left and right outputs of the stereo decoder.

Switch S3A3 determines the audio signal to be applied to the linear audio head. If SAP is not present, L+R is applied automatically. SAP is applied at pin 3 of IC3A4, while L+R is applied at pin 5 (from Q3D0/Q3D1). The switches within IC3A4 (for the linear audio head) are controlled by the logic at pin 9.

A high at pin 9 of IC3A4 connects the SAP input to the linear-head output at pin 4, while a low at pin 9 connects the L+R input to pin 4. When S3A3 is in the L+R position, pin 9 is grounded through S3A3. As a result, only L+R can be applied to the linear head. When S3A3 is in SAP, the logic at pin 9 of IC3A4 is determined by the $\overline{\text{SAP}}$ status command from the SAP signal-detect circuit (Sec. 4–2.11). The $\overline{\text{SAP}}$ status line is low when SAP is present, and high when SAP is not being transmitted.

When SAP is present, the collector of Q3017 is low. This low in inverted to a high by Q3H5 and applied to pin 9 of IC3A4. This connects pins 3 and 4 of IC3A4 and applies SAP to the linear-head amplifier.

When there is no SAP being transmitted, the collector of Q3017 is high, and the high is inverted to a low by Q3H5. With pin 9 of IC3A4 low, pins 4 and 5 are connected, and L + R is applied to the linear-head amplifier.

4–3. TYPICAL TEST/ADJUSTMENT PROCEDURES

All of the notes described in Sec. 1–7.6 apply to the following test/adjustment procedures. There are 10 adjustments required for the circuits described in Sec. 4–2. These include adjusting the L − R decoder and SAP detector PLLs so that the corresponding VCO is locked to the incoming signals, adjusting the stereo and SAP levels, adjusting the level of the SAP switch signal, adjusting L − R separation, and adjusting both the stereo and SAP dbx circuits for timing and variable deemphasis.

4–3.1 L − R decoder adjustment

Figure 4–20 is the adjustment diagram. The purpose of the L − R decoder IC3001 is to reinsert a carrier into the AM stereo L − R sidebands at pin 2 and produce

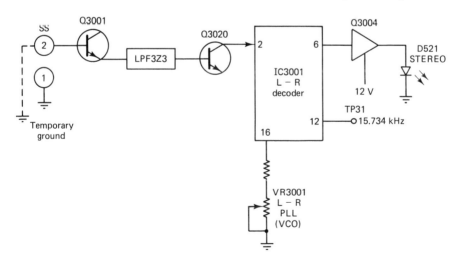

FIGURE 4–20 L−R decoder adjustment diagram.

corresponding audio output at pin 5. The missing L−R carrier is at 31.468 kHz and is produced by a VCO within the L−R decoder. Since there is no L−R carrier at the decoder output, the VCO is usually locked to the 15.734-kHz pilot (or a multiple).

In the circuit of Fig. 4−20, the VCO can be set by adjustment of PLL adjust VR3001 at pin 16 of IC3001 and can be monitored at TP31 (pin 12). The VCO signal at TP31 is 15.734 kHz, even though the VCO operates at a different frequency (typically higher).

Proceed as follows to adjust the L−R decoder.

1. Ground the input to the stereo decoder circuits at pin 2 of the SS connector. This removes all signals to the L−R decoder input, including the 15.734-kHz pilot. If the pilot is present, it is possible that the VCO will lock on to the incoming signal, even though the VCO is not exactly on frequency when free-running. Check that the STEREO indicator D521 is off.
2. Connect a frequency counter to TP31.
3. Adjust VR3001 for a reading of 15.734 kHz at TP31.
4. Remove the ground from pin 2 of the SS connector. Check that the TP31 reading remains at 15.734 kHz.

4–3.2 L−R gain adjustment

Figure 4−21 is the adjustment diagram. The purpose of this adjustment is to set the level of the stereo L−R signal before dbx noise-reduction processing. This is not to be confused with the L−R separation adjustment that sets the L−R level after dbx processing.

Proceed as follows to adjust the L−R level or gain.

1. Apply a modulated L−R signal to pin 2 of the SS connector. The STEREO indicator D521 should turn on. Use a modulation frequency of 300 Hz unless otherwise specified in the service literature.
2. Monitor TP37 for the 300-Hz signal, using an audio voltmeter or oscilloscope.
3. Adjust VR3002 for the correct voltage level at TP37. Always use the values specified in service literature (for both input amplitude at pin 2 of the SS connector and output at the test point).

4–3.3 SAP level adjustment

Figure 4−22 is the adjustment diagram. The purpose of this adjustment is to set the level of the SAP signal (before dbx processing) in relation to the stereo L−R signal. If the SAP and L−R signals are not approximately equal at this stage in the audio path, the user audio output or volume control must be reset each time when switching between stereo and SAP.

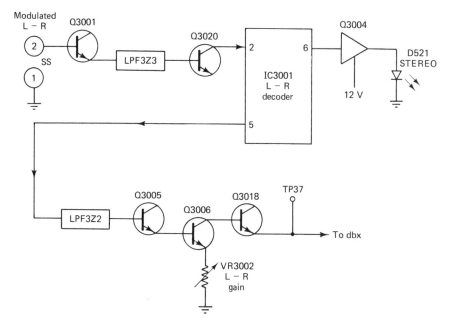

FIGURE 4-21 L – R gain adjustment diagram.

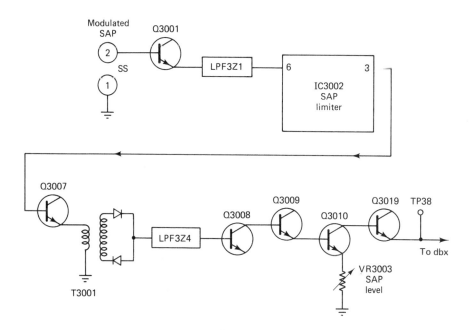

FIGURE 4-22 SAP level adjustment diagram.

Proceed as follows to adjust the SAP level or gain.

1. Apply a modulated SAP signal to pin 2 of the SS connector. The SAP indicator D520 (Fig. 4-18) should turn on. Use a modulation frequency of 300 Hz unless otherwise specified in the service literature.
2. Monitor TP38 for the 300-Hz signal, using an audio voltmeter or oscilloscope.
3. Adjust VR3003 for the correct voltage level at TP38. Always use the values specified in the service literature (for both the input amplitude at pin 2 of the SS connector and output at the test point).

Note that the SAP level adjustment is usually performed after adjustment of the SAP detector (but check the service literature for the preferred sequence).

4-3.4   SAP detector adjustment

Figure 4-23 is the adjustment diagram. The purpose of the SAP signal detector IC3003 is to produce a low at output pin 7 when the 78.67-kHz SAP carrier is present at input pin 6. (Output pin 7 remains high when the SAP carrier is not present.)

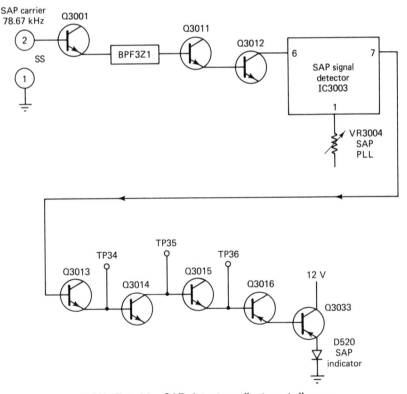

FIGURE 4-23   SAP detector adjustment diagram.

This is done by comparing the incoming SAP carrier with the IC3003 decoder VCO. When both signals are present and locked in frequency/phase, pin 7 goes low. In the circuit of Fig. 4–23, the VCO is set by adjustment of SAP PLL adjust VR3004 at pin 1 of IC3003.

Proceed as follows to adjust the SAP detector adjustment.

1. Apply a 78.67-kHz signal to pin 2 of the SS connector. Use the SAP carrier signal from the MTS generator described in Chapter 2.

2. Monitor the voltage across TP35 and TP36 as shown in Fig. 4–23.

3. Set VR3004 to full counterclockwise. The SAP indicator D520 should be off. The voltmeter should read about −1.0 V.

4. Adjust VR3004 clockwise until the voltage at TP35 and TP36 changes to about +1.0 V. This indicates that pin 7 of IC3003 is switched to low. You could measure at pin 7 of IC3003, or at TP345, but the change in signal level is much more difficult to detect. Also, by checking at TP35 and TP36, you also confirm operation of Q3013, Q3014, and Q3015 simultaneously.

5. With the SAP carrier still applied at the circuit input (pin 2 of the SS connector), check that the SAP indicator D520 has turned on when the TP35/TP36 voltage indication changes from −1.0 V to +1.0 V. This confirms operation of both D520 and Q3033.

Note that the SAP detector adjustment is usually performed after adjustment of the SAP detector signal output level. Check the service literature for the correct sequence of adjustments.

4–3.5 SAP detector signal output-level adjustment

Figure 4–24 is the adjustment diagram. The purpose of VR3005 is to set the level of the SAP detect signal. (This signal appears when the SAP carrier is present and pin 7 of IC3003 goes low.)

Proceed as follows to adjust the SAP detector signal output level.

1. Make certain that there is no SAP carrier signal present. (On some circuits it may be necessary to ground the SAP circuit input at pin 2 of connector SS. Check the service literature.)

2. Monitor the voltage across TP35 and TP36 as shown in Fig. 4–24.

3. Adjust VR3005 until the voltage at TP35 and TP36 is −1.0 V.

4–3.6 Stereo noise-reduction time-constant adjustment

Figure 4–25 is the adjustment diagram. The purpose of VR3006 is to set the *amount of control output* from dbx noise reduction IC3004 for a given L−R signal amplitude.

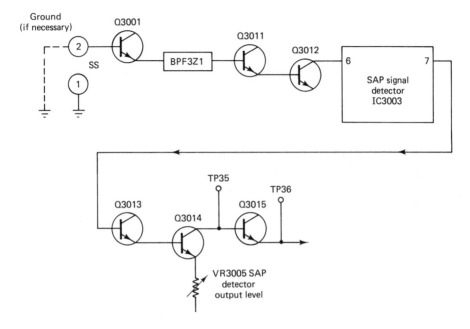

FIGURE 4-24 SAP detector output-level adjustment diagram.

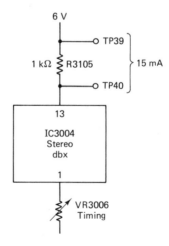

FIGURE 4-25 Stereo noise-reduction time-constant adjustment diagram.

Proceed as follows to adjust the stereo NR time constant.

1. Connect a digital multimeter to TP39 and TP40 as shown.
2. Adjust VR3006 for the correct voltage level across R3105 (at TP39 and TP40). Use the service literature values.

Note that although you are measuring voltage across R3105, VR3006 is set for a given current through circuits within IC3004 (the more current, the more control). Typically, VR3006 is set for a current of 15 mA at pin 13 of IC3004. Since R3105 is 1 kΩ, the voltage reading should be 15 mV.

4-3.7 Stereo noise-reduction VD adjustment

Figure 4–26 is the adjustment diagram. The purpose of VR3007 is to set the L – R gain, after spectral processing by IC3004 but before wideband processing, to produce the desired variable deemphasis or VD. (This is sometimes called the *wideband* or *high-band* VD adjustment.)

Proceed as follows to adjust the stereo VD.

1. Make certain that there is no L – R signal present. On some circuits this can be done by grounding the input at pin 2 of the SS connector. In other circuits it is necessary to disable the L – R audio path (such as connecting the emitter of Q3018 to +9 or +12 V).
2. Apply a 300-Hz sine wave to TP37, using the correct level specified in the service literature. A typical input level at TP37 is – 24 dB.
3. Monitor the 300-Hz signal (after noise reduction) at TP43. Make certain that the level at TP43 is within limits specified by the service literature. The typical 300-Hz output level at TP43 should be between – 23 and – 35 dB. Note the actual level at TP43.

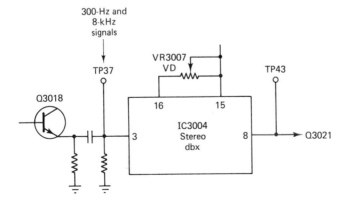

FIGURE 4–26 Stereo noise-reduction VD adjustment diagram.

4. Change the frequency of the audio signal applied at TP37 from 300 Hz to 8 kHz. Set the amplitude of the 8-kHz signal as specified in service literature. A typical input level at TP37 (with 8-kHz audio) is −17 dB.

5. Adjust VR3007 so that the 8-kHz signal (after noise reduction) at TP43 is as specified in the service literature. A typical 8-kHz output level at TP43 is the actual 300-Hz level (measured in step 3) less −11 dB.

Keep in mind that these noise-reduction adjustments are critical to proper operation of the MTS circuits. Also, the service literature generally recommends that the VD adjustment described here, and the time-constant adjustment described in Sec. 4–3.6, be performed before the separation adjustment of Sec. 4–3.10.

4–3.8   SAP noise-reduction time-constant adjustment

Figure 4–27 is the adjustment diagram. The purpose of VR3008 is to set the *amount of control output* from dbx noise reduction IC3006 for a given SAP signal amplitude. Proceed as follows to adjust the SAP NR time constant.

1. Connect a digital multimeter to TP41 and TP42 as shown.

2. Adjust VR3008 for the correct voltage level across R3126 (at TP41 and TP42). Use the service literature values.

Note that although you are measuring voltage across R3126, VR3008 is set for a given current through circuits within IC3006 (the more current, the more control). Typically, VR3008 is set for a current of 15 mA at pin 13 of IC3006. Since R3126 is 1 kΩ, the voltage reading should be 15 mV.

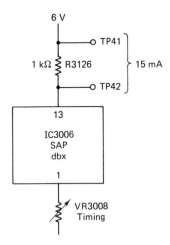

FIGURE 4–27   SAP noise-reduction time-constant adjustment diagram.

4–3.9 SAP noise-reduction VD adjustment

Figure 4–28 is the adjustment diagram. The purpose of VR3009 is to set the SAP gain, after spectral processing by IC3006 but before wideband processing, to produce the desired variable deemphasis or VD. (This is sometimes called the *wideband* or *high-band* VD adjustment.)
Proceed as follows to adjust the SAP VD.

1. Make certain that there is no SAP signal present. On some circuits, this can be done by grounding the input at pin 2 of the SS connector. In other circuits, it is necessary to disable the SAP audio path (such as connecting the emitter of Q3019 to +9 or +12 V).

2. Apply a 300-Hz sine wave to TP38, using the correct level specified in the service literature. A typical input level at TP38 is −24 dB.

3. Monitor the 300-Hz signal (after noise reduction) at TP44. Make certain that the level at TP44 is within the limits specified by the service literature. The typical 300-Hz output level at TP44 should be between −23 and −35 dB. Note the actual level at TP44.

4. Change the frequency of the audio signal applied at TP38 from 300 Hz to 8 kHz. Set the amplitude of the 8-kHz signal as specified in the service literature. A typical input level at TP38 (with 8-kHz audio) is −17 dB.

5. Adjust VR3009 so that the 8-kHz signal (after noise reduction) at TP44 is as specified in the service literature. A typical 8-kHz output level at TP43 is the actual 300-Hz level (measured in step 3) less −11 dB.

Keep in mind that these noise-reduction adjustments are critical to proper operation of the MTS circuits (to the SAP audio in this case).

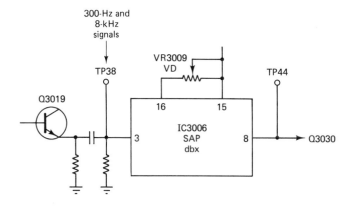

FIGURE 4–28 SAP noise-reduction VD adjustment diagram.

4–3.10 L–R separation adjustment

Figure 4–29 is the adjustment diagram. The purpose of this adjustment is to set the level of the stereo L – R signal in relation to the mono L + R signal. Both signals are combined in the Q3022 matrix (R3159/R3160). If the L – R signal is low in relation to L + R, you will hear only mono. If the L – R is high in relation to L + R, you will hear both signals, but there will be poor separation between left and right audio (the audio will sound like mono, even though stereo is present). Proceed as follows to adjust the L – R separation.

1. Apply a modulated L – R signal to pin 2 of the SS connector. The STEREO indicator D521 should turn on. Use the modulated signal from the MTS generator described in Chapter 2. Use a modulation frequency of 300 Hz unless otherwise specified in the service literature. Refer to Sec. 2–6.9 for further information on the separation adjustment.

2. Monitor TP32 and TP33 for the 300-Hz signal, using an audio voltmeter.

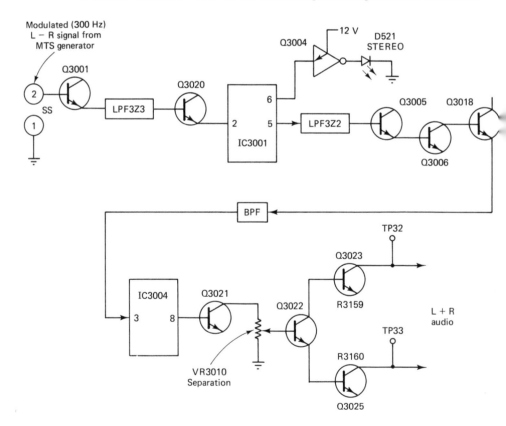

FIGURE 4–29 L – R separation adjustment diagram.

3. Adjust VR3010 for the *correct voltage level* at TP32 and TP33. Always use the values specified in service literature (for both input amplitude and output at the test points). Keep in mind that this adjustment can be *critical* in producing good stereo sound.

4. If you cannot find a separation adjustment procedure in the service literature, use the following as an *emergency procedure only*.

5. Set VR3010 to the full counterclockwise position, so that L – R is zero. Apply an L + R signal with 300-Hz modulation at pin 2 of the SS connector. Note the voltmeter reading at TP32 and TP33. This is the mono L + R signal.

6. Remove the L + R signal and apply L – R with 300-Hz modulation at pin 2 of the SS connector.

7. Adjust VR3010 until the reading at TP32 and TP33 are the same as with L + R (or just below L + R) in step 5.

8. For a final test, measure stereo separation as described in Sec. 2–6.7. (In fact, some service technicians recommend adjusting VR3010 to get a given separation, or maximum separation, at the loudspeakers, rather than for a given reading in the decoder circuits.) A stereo separation of 60 dB, or better, is possible on some VCRs.

4–4. TROUBLESHOOTING APPROACH

All of the notes described in Secs. 1–7 and 1–8 apply to the following procedures. In each of the following trouble symptoms, it is assumed that an active TV channel has been tuned in, that both stereo and SAP signals are (supposedly) being broadcast (in addition to a mono broadcast), and that video can be recorded and played back. If you do not have a good picture, the problem is likely in sections ahead of the MTS circuits (such as the RF, IF, or 4.5-Mhz audio carrier circuits). It is also possible that the problem is in the TV set used to monitor the VCR output.

As is the case with any VCR, a good preliminary troubleshooting step is to play back a known-good tape using a known-good TV set. As an alternative to a TV set, you can monitor the audio output of the VCR at the headphone jack and/or audio output jacks.

When checking MTS audio, the tape should have both hi-fi audio (recorded on the rotating-head track) and normal audio (recorded on the linear-head track).

If the audio playback is good on both tracks, it is reasonable to assume that the heads and playback amplifiers are good (as are any audio amplifiers that follow the playback amps).

If playback is good from one track but not the other, you have isolated the problem to either the hi-fi heads and amplifiers or the normal-audio head and amplifiers.

If there is no playback from either the hi-fi or normal tracks, or if playback is abnormal from both tracks, the problem is probably in the playback amps or audio amps following the MTS circuits.

Once you have established that audio playback is good, the next step is to record and play back broadcast audio. Check if any audio is available at the audio output jacks (or headphone jack), indicating that a mono broadcast is available. Then check if the STEREO and SAP indicators are turned on, indicating stereo and SAP broadcast signals.

4–4.1 No audio available at the audio output jacks

If there is no audio available at the audio output jacks, but the good picture is good and either the STEREO or SAP indicator is on, the problem is probably in the mono L + R audio path or in the audio amplifiers that follow the MTS circuits. The fact that either indicator is on makes it likely (but not absolutely certain) that the sections ahead of the MTS circuits are good.

Before you plunge into any circuits, make certain that the INPUT SELECT, AUDIO MONITOR, and NORMAL AUDIO switches are in the correct positions. For example, if the SAP indicator is on, do not expect to record SAP unless the NORMAL AUDIO switch is set to SAP.

There are several approaches that can be used at this point. Generally, the simplest is to inject a composite signal (with L + R) at the input of the MTS circuits. As shown in Fig. 4–15, this input is at pin 2 of the SS connector.

As an alternative, you can inject audio at the output of the MTS circuits (input to the audio circuits following the MTS circuits). If it is possible to record with audio applied at this point, but not through the MTS circuits (with a composite signal), the problem is probably in the MTS circuits.

If the modulated signal is not heard at the audio output jacks with an L + R signal applied, trace through the audio path using signal injection or signal tracing, whichever is most convenient. The L + R path (up to the audio input-select circuits) includes Q3001, LPF3Z3, Q3020, LPF3Z1, Q3002, Q3003, IC3008, Q3023, Q3024, Q3025, and Q3026, as shown in Fig. 4–15.

If the modulated signal is heard at the audio output jacks with composite L + R applied, it is reasonable to assume that the audio path through the MTS circuits and audio amplifiers is good. The problem is probably ahead of the MTS circuits (VCR tuner, IF, etc.), but could be in the audio input-select circuits.

4–4.2 No stereo operation, mono operation is good

If there is mono audio available at the audio output jacks, and the picture is good but there is no stereo operation, check that the STEREO indicator D521 is on (indicating that stereo is being broadcast).

If STEREO is on, the problem is probably in the L − R audio path shown in Fig. 4–15. If STEREO is off, the problem may be that no stereo is being broadcast! The quickest approach at this point is to inject a composite signal (with L − R) at the input of the MTS circuits (pin 2 of SS).

If STEREO turns on and you get stereo at the audio output jacks, the problem is probably ahead of the MTS circuits. Proceed as described in Sec. 4–4.1. If STEREO does not turn on and there is no stereo at the audio output jacks, the problem is in the L–R audio path. (The same is true if STEREO is on but there is no stereo at the audio output jacks.)

There are several points to consider when tracing through the L–R path, whether signal injection or signal tracing is used. First, try correcting the problem by adjustment of the L–R path as described in Sec. 4–3. These adjustments include the L–R decoder (PLL) VR3001, L–R gain VR3002, stereo dbx time constant VR3006, stereo dbx variable deemphasis VR3007, and L–R separation VR3010. These adjustments could cure the problem. If not, simply making the adjustments might lead to the defect.

For example, if there is no 15.734-kHz signal at TP33, no matter how VR3001 is adjusted, IC3001 is suspect. Similarly, if pin 6 of IC3001 does not go low when a pilot is injected at the MTS input, IC3001 is suspect.

4–4.3 No SAP operation, mono and stereo operation are good

If there is mono and stereo operation, and the good picture is good but there is no SAP operation, check that the NORMAL AUDIO switch S3A3 is in SAP and that the SAP indicator D520 is on (indicating that SAP is being broadcast).

If SAP is off, the problem may be that no SAP is being broadcast! Similarly, if NORMAL AUDIO is set to L+R, SAP will not be recorded!

If SAP is on and NORMAL AUDIO is set to SAP, the problem is probably in the SAP audio path (Fig. 4–17), SAP signal-detect path (Fig. 4–18), or possibly the audio input-select circuit (Fig. 4–19). The quickest approach at this point is to inject a composite signal (with SAP) at the input of the MTS circuits (pin 2 of SS).

If SAP turns on and you get SAP at pins 3 and 4 of IC3A4, the problem is probably ahead of the MTS circuits. Proceed as described in Sec. 4–4.1.

If SAP does not turn on and there is no SAP at IC3A4, the problem is in the SAP audio or signal-detect paths. (The same is true if SAP is on but there is no SAP at IC3A4.)

There are several points to consider when tracing through the SAP paths, whether signal injection or signal tracing is used. First, try correcting problems in the SAP paths by adjustment as described in Sec. 4–3. These adjustments include the SAP level VR3003, SAP detector (PLL) VR3004, SAP detector output level VR3005, SAP dbx time constant VR3008, and SAP dbx VD VR3009. These adjustments could cure the problem. If not, simply making the adjustments might lead to the defect.

For example, if pin 7 of IC3003 does not go low when there is an SAP signal applied at the input (Fig. 4–18), no matter how VR3004 is adjusted, IC3003 is suspect. (IC3003 is also suspect if pin 7 remains low when the SAP input is removed.)

If pin 7 of IC3003 does go low when SAP is applied, check that the high appears on the base of Q3033. If so, and SAP indicator D520 does not turn on,

D520 and Q3033 are suspect. If the base of Q3033 does not go high when there is an SAP input (and pin 7 of IC3003 is low), suspect Q3013, Q3014, Q3015, and Q3016. It is also possible that VR3005 is not properly adjusted. Refer to Sec. 4-3.5.

If the SAP indicator D520 is turned on, check for a high at pin 9 of IC3A4 (Fig. 4-19). If pin 9 is high, check that the audio at pins 3 and 4 of IC3A4 is substantially the same. If pin 9 is not high, suspect Q3H5 and Q3017.

If there is no audio at pin 3 of IC3A4, trace through the SAP audio path (Fig. 4-17). Keep in mind that pin 3 of IC3A4 should be high when SAP is present. When SAP is removed, the output of Q3017 goes high. This high is inverted by Q3032 to allow at pin 3 of IC3A4 (to mute the SAP audio). The high is also inverted by Q3045 to a low at the base of Q3008 (also to mute the SAP audio).

4-4.4 Audio input-switching problems

Problems in audio input-switching circuits (Fig. 4-19) can produce trouble symptoms that appear to be in other circuits. For example, if there is no mono or stereo, it is possible that IC3A4 has not received a command that selects pins 1 and 13. This could be the fault of S3A1 or the associated circuit. Similarly, IC3A4 may receive the correct command but not respond properly. This is the fault of IC3A4.

Before you condemn the MTS circuits ahead of the audio input switching, check all signals to and from IC3A4. As an example, set S3A1 to TV and check that pins 9, 10, and 11 are high. If not, suspect S3A1, R3F8, and R3G8.

If pins 10 and 11 are high, check that the signal at pin 14 of IC3A4 is substantially the same as at pin 13, while the signal at pin 15 is the same as at pin 1. If not, suspect IC3A4. If pin 9 is high, check that the signal at pin 4 is substantially the same as at pin 3. If not, suspect IC3A4.

Now set S3A3 to L + R and check that pin 9 of IC3A4 is low (no matter what the position of S3A1). If not, suspect S3A3. If pin 9 is low, check that the signal at pin 4 is substantially the same as at pin 5. If not, suspect IC3A4.

Note that the L + R signal is applied to IC3A4 through Q3D0 and Q3D1. So if it is possible to record and play back SAP on the linear head but not L + R (even though pin 9 of IC3A4 is low and pin 4 is connected to pin 5), suspect Q3D0 and Q3D1.

5

SONY MTS
ADAPTER

This chapter is devoted to the circuits of a typical Sony MTS adapter. The adapter makes it possible to receive both stereo and SAP broadcasts on a monaural TV set.

If the TV set has a built-in stereo amplifier and loudspeakers, the stereo/SAP can be played through the built-in system. If the TV set has only one loudspeaker, the adapter can be used with external loudspeakers and/or stereo equipment.

Note that the TV set must have an MPX output from the tuner (as do Sony Profeel tuners and Trinatron TV sets), as well as audio inputs (unless external audio equipment is used).

The required MPX input is 77.5 mV rms, with a load impedance of 10 kΩ, sync positive.

The LINE IN and LINE OUT signals are −5 dB (435 mV rms, where 0 db = 0.775 V rms). The LINE IN load impedance is more than 47 kΩ, while the LINE OUT load impedance is 4.7 kΩ.

The HEADPHONES jack (stereo minijack) accepts low- and high-impedance stereo headphones.

The SPEAKER L/R terminals can be connected to loudspeakers in the range 4 to 8 Ω.

5-1. TYPICAL ADAPTER CONNECTIONS

Figure 5-1 shows some typical connections between the adapter and TV set, VCR, or stereo equipment. Always follow the connection diagrams (and notes) in the service literature. In the absence of specific instructions, here are some notes that apply to most MTS adapters.

Turn off *all components* before making any connection.

Make connections firmly. A loose connection can produce distorted sound.

Connect the ac power cord *after all* of the other connections are completed.

When using the adapter with a stereo-ready TV set (with MPX output and stereo audio inputs), regulate the sound level with the TV volume control.

When using the adapter with a TV tuner, regulate the sound level with the tuner volume control.

When using the adapter with a mono TV and external loudspeakers, regulate the sound level and the adapter volume control.

When connecting to a VCR, the VCR should be a hi-fi model. If the TV is monaural, connect a pair of stereo loudspeakers to the adapter.

You can record stereo sound of the MTS programs on a stereo cassette deck or recorder through the adapter, and then listen to the recorded sound (again through the adapter). The adapter can be connected directly to an existing stereo system. In this case, you can regulate the sound level with both stereo controls and adapter volume control.

5-2. OPERATING (USER) CONTROLS AND INDICATORS

Figure 5-2 shows the operating controls and indicators for the adapter. The following paragraphs describe the control and indicator functions.

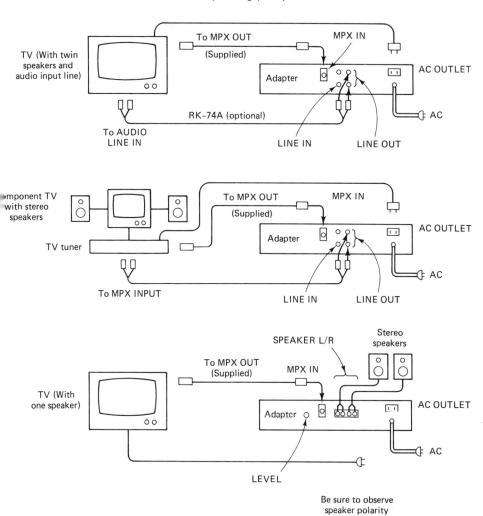

FIGURE 5–1   Typical connections between the adapter and TV set.

Press the MAIN button to select the main-channel sound. The MAIN lamp should turn on. When a stereo broadcast is being received, as indicated when the STEREO lamp turns on, press MAIN to select stereo.

Press the SAP button to select SAP sound. The SAP lamp should turn on.

Press the BOTH button to select both main-channel and SAP sound. Both the MAIN and SAP lamps should turn on. The stereo broadcast of the main-channel becomes mono (L + R) when BOTH is pressed, and the main-channel sound is heard only on the left loudspeaker.

Press the AUTO STEREO button. The STEREO lamp should turn on when stereo is being broadcast.

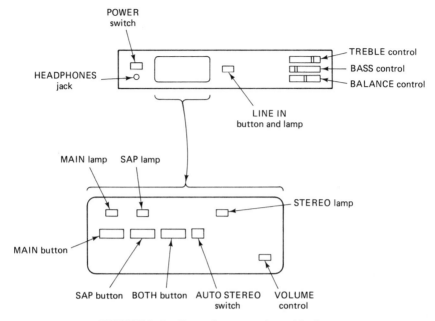

FIGURE 5–2   Operating controls and indicators.

The BALANCE control regulates the amount of sound coming from the left and right loudspeakers. Adjust BALANCE for best stereo effect (or set BALANCE at the center detent position).

Slide the TREBLE and BASS controls for best audio response (or set both at the center position). Slide towards the right (+) to increase the bass/treble, and to the left (−) to decrease effect.

Connect stereo headphones to the HEADPHONE jack (stereo minijack).

Press the POWER button to turn on the adapter. Press POWER again to turn the adapter off.

Press the LINE IN button to select sound from an alternative audio source connected to the rear-panel LINE IN jacks. The LINE IN lamp should turn on. Usually, LINE IN is left in the released position (to select TV sound).

Set the VOLUME control to the desired audio level (from both the HEAD-PHONES jack and rear-panel SPEAKER L/R terminals). Slide toward the right to increase volume.

Set the rear-panel LEVEL control to the center detent position. When the incoming TV signal is weak and the stereo or SAP broadcast is not being received properly, turn the LEVEL control clockwise to increase the broadcast signal level.

5–2.1 Typical operating sequence

Figure 5–3 shows the typical operating sequence for the adapter. The following notes supplement the illustration.

To receive a stereo program, press AUTO STEREO. Under these conditions, the

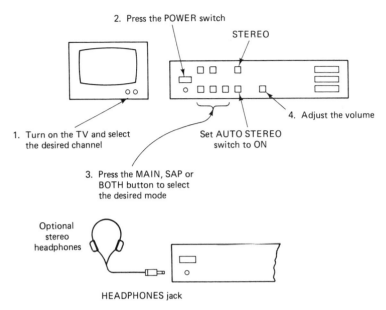

FIGURE 5-3 Typical operating sequence.

STEREO lamp turns on when a stereo program is broadcast. To hear the program in stereo, press MAIN and check that the MAIN lamp turns on. With MAIN pressed, the stereo left and right channels are heard on the respective loudspeakers.

To receive SAP (when available), press SAP. The SAP lamp should turn on and SAP should be heard from both loudspeakers.

To receive both main-channel and SAP signals (when both are available), press BOTH. Both the MAIN and SAP lamps should turn on. The main-channel left and right channels are combined and heard on the left loudspeaker (as mono L+R sound). The SAP channel is heard on the right loudspeaker.

Note that this same channel distribution takes place at the rear panel LINE OUT terminals. This makes it possible to record both channels on a stereo VCR. The choice of programs can then be selected during playback.

Also note that there may be stereo broadcasts where excessive noise is heard (due to a weak broadcast signal). In some cases this problem can be cured by setting the AUTO STEREO switch to off and returning to a mono mode. It may also be possible to cure the problem by setting the rear-panel LEVEL control clockwise.

Keep in mind that there may be problems ahead of the adapter circuits (bad antenna, weak TV tuner, etc.) if it is necessary to set the LEVEL control full clockwise for all TV channels, or if stereo cannot be received on any channel.

5-3. CIRCUIT DESCRIPTIONS

All of the notes described in Sec. 1-7 apply to the following descriptions. Note that the adapter circuits perform the same functions (decoding, dbx processing, etc.) as those described for MTS circuits in other chapters of this book (in addition to providing some audio amplification). However, the L+R, L−R, and SAP audio paths are quite different.

5–3.1 MPX input and SAP audio path

Figure 5–4 shows the MPX (or composite signal input and SAP audio path in simplified form. The MPX input from the tuner/IF is applied to the adapter circuit through rear-panel LEVEL control RV905. The composite signal is amplified by

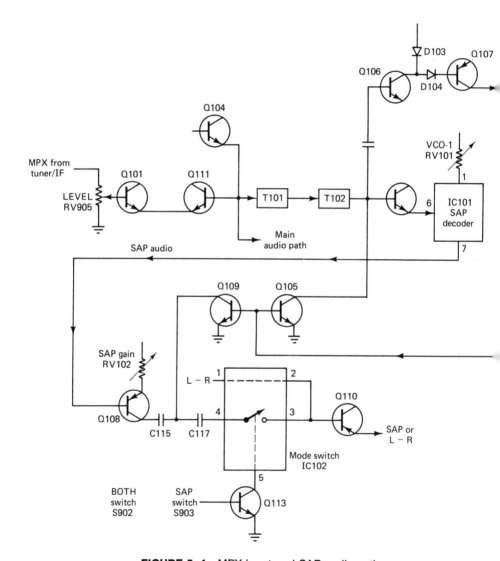

FIGURE 5–4   MPX input and SAP audio path.

Q101, Q111, and Q104. The output from buffer Q104 is applied to the main and SAP audio paths through filters.

The SAP signal is passed by filters T101/T102 and is applied to SAP decoder IC101 through amplifier Q103. SAP audio is taken from pin 7 of IC101 and applied to mode switch IC102 through amplifier Q108.

VCO-1 control RV101 sets the frequency of the VCO within IC101 (78.67 kHz). Gain or level control RV102 sets the gain or level of SAP amplifier Q108.

A SAP-mute circuit consisting of Q105, Q106, Q107, and Q109 disables the SAP audio path when there is no SAP carrier signal.

The SAP signal is amplified by Q106, rectified by D103/D104, and applied to Q107. If SAP is present, the base of Q107 is high (about 14 V) and the collector is zero. This keeps Q105/Q109 off.

If SAP is not being broadcast, the base of Q107 goes low (about 8 V) and the collector rises to about 9 V. This drives both Q105 and Q109 into conduction, bypassing any signals in the SAP path to ground.

The SAP output from mode switch IC102 at pin 3 is applied to amplifier Q110, along with the L−R signal (Sec. 5-3.2). The SAP signal at pin 3 of IC102 is controlled by Q113, which, in turn, is controlled by SAP switch S903 or BOTH switch S902.

When either SAP or BOTH is selected, the base of Q113 goes low (zero) and the collector rises to 8 V, connecting pins 3 and 4 of IC102. When SAP is off, the base of Q113 goes high (about 0.7 V) and the collector drops to zero, disconnecting pins 3 and 4 of IC102.

5–3.2 Main (L+R) and L−R audio paths

Figure 5-5 shows the L+R and L−R audio paths in simplified form. The L+R and L−R signals are passed by filter T103 and applied to main decoder IC106 through buffer Q102. The signals at pins 5 and 6 of IC106 are applied to amplifiers within IC105 to become the L+R and L−R audio.

L+R audio is applied to the mixer or matrix circuit (Sec. 5-3.4) through IC105, T104, and Q120. Gain or level control RV106 sets the gain or level of L+R amplifier Q120.

L−R audio is applied to the mode switch IC102 through IC105. L−R control RV103 sets the gain or level of the L−R audio. VCO 2 control RV104 sets the frequency of the L−R VCO within IC106 (62.936 kHz).

The L−R output from mode switch IC102 at pin 2 is applied to amplifier Q110, along with the SAP signal (Sec. 5-3.1). The L−R signal at pin 2 of IC102 is controlled by Q112, which, in turn, is controlled by MAIN switch S904, BOTH switch S902, and SAP switch S903.

When MAIN is selected, the base of Q112 goes low (zero), and the collector rises to 9 V, connecting pins 1 and 2 of IC102. When MAIN is off, the base of Q112 goes high (about 0.6 V) and the collector drops to zero, disconnecting pins 1 and

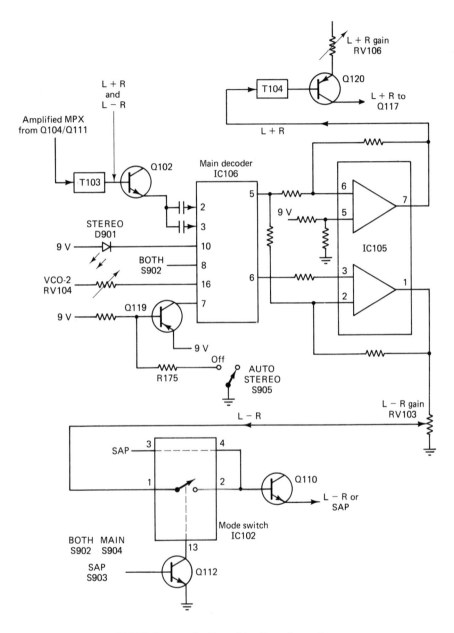

FIGURE 5–5 L+R and L–R audio paths.

2 of IC102. The choice between mono (L + R) or stereo (L − R) is determined by the Q119 signal applied at pin 7 of IC106.

When AUTO STEREO switch S905 is set to off, the base of Q119 goes low. This drives Q119 into conduction and applies a high to IC106. Under these conditions, the L − R VCO is disabled, only L + R passes, and pin 10 of IC106 goes high (about 8 V), turning off STEREO lamp D901.

When AUTO STEREO switch S905 is pressed (on), the base of Q119 is floating (about 9 V) and Q119 remains off, dropping pin 7 of IC106 to about 3.8 V. Under these conditions, the L − R VCO is turned on, both L + R and L − R pass, and pin 10 of IC106 goes low (zero), turning on STEREO lamp D901.

Note that pin 10 of IC106 does not go low unless there is a pilot signal present. So D901 may not turn on even though AUTO STEREO has been selected.

When BOTH is selected, pin 8 of IC106 goes low, disabling pin 6. Under these conditions, only L + R is available at IC105.

5–3.3 dbx noise-reduction path

Figure 5–6 shows the dbx noise-reduction path in simplified form. Note that both SAP and L − R use the same path. The SAP/L − R signals are applied to dbx noise reduction IC501 through Q110, T105, Q122, Q501, and Q502.

SAP/L − R audio from the emitter of Q502 is applied to the spectral VCA at pin 18 of IC501 through a fixed deemphasis network. Audio signals from the emitter of Q122 (in the range 100 Hz to 3 kHz) are applied to the wideband rms detector at pin 3 of IC501. C505/C506 and R504/R505 form the wideband BPF. Audio from the emitter of Q501 (in the range 4 to 15 kHz) are applied to the spectral rms detector at pin 20 of IC501. C509/C510 and R512/R513/R514 form the spectral BPF.

Two rms detectors in IC501 receive power from a constant-current generator. The output of this generator is adjustable by RV501, the timing control, connected at pin 1 of IC501. The setting of RV501 determines the amount of control output from the rms detectors for a given signal amplitude.

Spectral processing is performed by the spectral VCA at pin 18 of IC501, controlled by the output of the spectral rms detector. The signal is then applied to an amplifier within IC502 through C513/C515 and an op-amp within IC501. VD control RV502 sets the signal level after spectral processing, but before wideband processing, by setting the gain of the op-amp.

The amplified signal is then applied to the wideband VCA at pin 5 of IC501. Wideband processing is performed by the wideband VCA, controlled by the output of the wideband rms detector.

The processed signal is then applied to another op-amp within IC501 through C522. The op-amp output at pin 8 of IC501 is applied to the matrix and audio-output circuit (Sec. 5–3.4) through another amplifier in IC502. Separation control RV503 sets the gain or level of the processed signal (and thus the separation) when L − R is being processed. The network at pins 7 and 8 of IC501 produce the required fixed deemphasis.

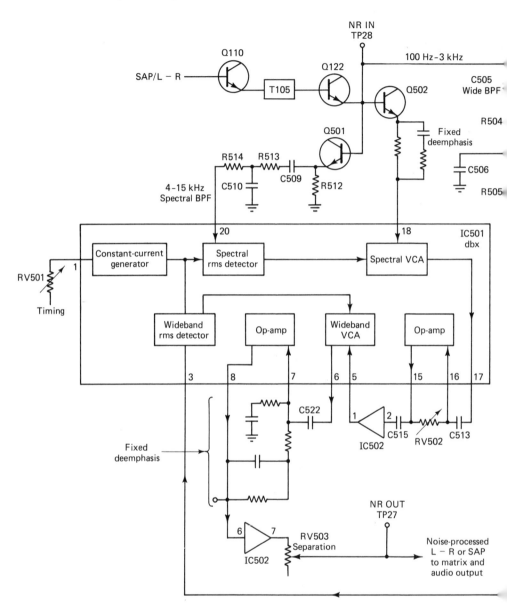

FIGURE 5-6   dbx noise-reduction path.

5-3.4 Matrix and audio output path

Figure 5-7 shows the matrix and audio output path in simplified form. The L + R, L – R, and SAP signals are applied to audio control IC107 through mode switch IC102, mode switch IC104, and the associated matrix circuits (Q116, Q117, and Q118).

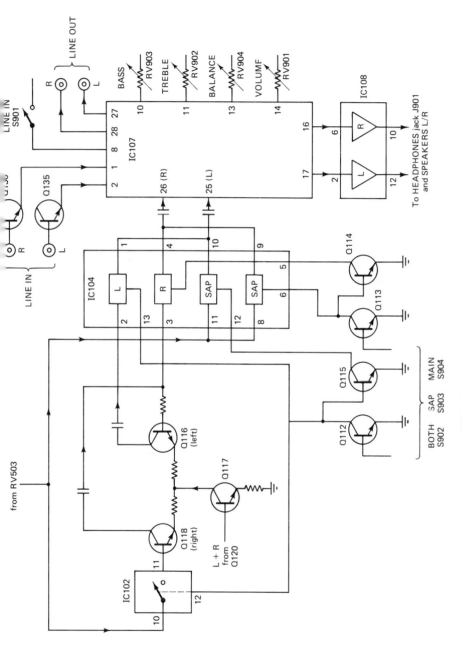

FIGURE 5-7 Matrix and audio output path.

The L + R signal from the emitter of Q120 is applied to the base of Q117. The SAP/L − R signal from RV503 is applied to pins 8 and 11 of IC104 and to pin 10 of IC102.

When MAIN is selected and there is no stereo and/or when AUTO STEREO is off, only the L + R signal from Q117 is applied to pins 25 and 26 of IC107 through Q116 (left), Q118 (right), and IC104 (pins 1/2 left, pins 3/4 right). This produces a monaural input to IC107 and mono output to the SPEAKERS L/R terminals, HEADPHONES jack, and LINE OUT jacks.

When MAIN is selected, there is stereo being broadcast, and AUTO STEREO is on, L − R from RV503 (through IC102 and Q118) is combined with L + R from Q117 to produce a stereo input to IC107 (pin 25 left, pin 26 right). This results in a stereo output to the loudspeakers, headphones, and LINE OUT.

When SAP is selected, SAP (if any) from RV503 is applied to IC107 through IC104 (pins 10/11 and 8/9) to produce a monaural SAP output to the loudspeakers, headphones, and LINE OUT. L + R signals are disconnected from IC107 since pins 1/2 and 3/4 of IC104 are open.

When BOTH is selected, Q113 is turned off by S902, S903, and S904. The collector of Q113 rises to about 9 V, as does pin 6 of IC104. The SAP signal from RV503 is applied through pins 8 and 9 of IC104 to pin 26 of IC107, producing a SAP output on the right channel.

When Q113 turns off, Q114 turns on and pin 5 of IC104 goes low, opening pins 1/2 of IC104 and preventing L + R at pin 2 from passing to the right channel.

When BOTH is selected, Q112 is turned off by S902, S903, and S904. The collector of Q112 rises to about 9 V, pin 13 of IC104. The L + R signal from Q117 is applied through Q116 and pins 1/2 of IC104 to pin 25 of IC107, producing an L + R output on the left channel.

When Q112 turns off, Q115 turns on and pin 12 of IC104 goes low, opening pins 10/11 and preventing SAP at pin 11 from passing to the left channel.

Audio signals to and from the adapter are passed through IC107. In turn, IC107 is controlled by LINE IN switch S901.

When S901 is pressed, the LINE IN input is applied to the LINE OUT jacks through Q130/Q135 and IC107. Also, the LINE IN input is applied to the HEADPHONES jack J901 and SPEAKER L/R terminals through Q130/Q135, IC107, and IC108.

When S901 is released, the L + R, L − R, and SAP audio at pins 25 and 26 of IC107 are applied to the LINE OUT jacks, HEADPHONES jack J901, and SPEAKER L/R terminals through IC107 and IC108. IC107 also provides a means of controlling volume, balance, treble, and bass for the audio signals, through corresponding controls.

5–4. TYPICAL TEST/ADJUSTMENT PROCEDURES

All of the notes described in Sec. 1-7.6 apply to the following test/adjustment procedures.

5–4.1   Preliminary setup and signal requirements

For all of the following test/adjustment procedures, the rear-panel LEVEL control RV905 (Fig. 5–4) should be set to the center detent position, and the input signal levels should be measured at RV905.

Note that the test/adjustment input signal levels are not necessarily those obtained directly from the MTS generator described in Chapter 2. However, it is possible to produce the required signal levels and frequencies using the external modulation connections of the MTS generator. Always use the input signal values specified in the service literature.

The required test/adjustment signals are as follows:

L+R: 300 Hz, 77.5 mV rms (0.22 V p-p)

L+R: 400 Hz, 100% modulation carrier, 31.468 kHz, 112 mV rms (0.44 V p-p)

SAP: 400 Hz, 100% modulation (±10 kHz deviation), 46.5 mV rms (0.13 V p-p)

Stereo pilot: 15.734 kHz, 15.6 mV rms (0.044 V p-p)

5–4.2   SAP decoder VCO-1 adjustment

Figure 5–8 is the adjustment diagram. The purpose of the SAP decoder IC101 is to produce audio at pin 7 when there is a SAP carrier and audio at input pin 6.

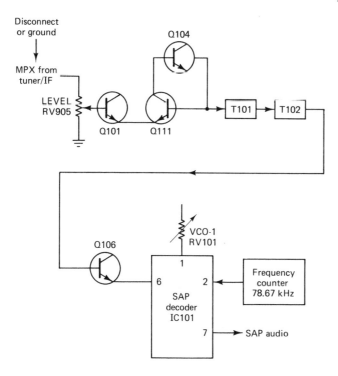

FIGURE 5–8   SAP decoder VCO-1 adjustment diagram.

IC101 contains a VCO operating at 78.67 kHz (the SAP carrier frequency) that must be set when free-running. The VCO can be set by VCO-1 control RV101 and can be monitored at pin 2 of IC101.
Proceed as follows to adjust the SAP decoder.

1. Disconnect the MPX input to the adapter (or ground the MPX input). This eliminates any possibility of a SAP carrier or pilot entering the circuits. If the carrier or pilot is present, it is possible that the VCO will lock on to the incoming signal, even though the VCO is not exactly on-frequency when free-running.
2. Connect a frequency counter to pin 2 of IC101.
3. Adjust RV101 for a reading of 78.67 kHz ±300 Hz.
4. Replace the MPX connector from the tuner/IF (and/or remove the MPX input ground). Check that the reading at pin 2 of IC101 remains at 78.67 kHz.

5–4.3   Stereo decoder VCO-2 adjustment

Figure 5–9 is the adjustment diagram. One purpose of main decoder IC106 is to reinsert a carrier into the AM stereo L−R sidebands and produce corresponding audio output at pins 5 and 6. The missing L−R carrier is at 31.468 kHz and is

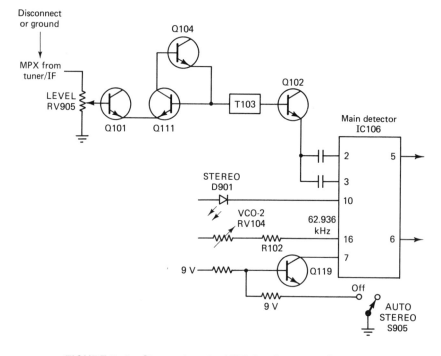

FIGURE 5–9   Stereo decoder VCO-2 adjustment diagram.

produced by a VCO within IC106. Since there is no L−R carrier available to IC106, the VCO is usually locked to the 15.734-kHz pilot (or a multiple).
In the circuit of Fig. 5–9, the VCO is set by VCO-2 control RV104 at pin 16 of IC106, and can be monitored at R201 (also at pin 16). The VCO signal at R201 is 62.936 kHz, even though the L−R requires a carrier of 31.468 kHz.
Proceed as follows to adjust the stereo decoder.

1. Disconnect the MPX input to the adapter (or ground the MPX input). This removes all signals to IC106, including the 15.734-kHz pilot. If the pilot (or multiple) is present, it is possible that the VCO will lock on to the incoming signal, even though the VCO is not exactly on-frequency when free-running.
2. Press the AUTO STEREO switch S905. This enables the VCO in IC106. However, STEREO lamp D901 should not turn on since there is no pilot.
3. Connect a frequency counter to R201.
4. Adjust RV104 for a reading of 62.936 kHz ±300 Hz.
5. Replace the MPX connector from the tuner/IF (and/or remove the MPX input ground). Check that the reading at R201 remains at 62.936 kHz.

5–4.4 L+R gain adjustment

Figure 5–10 is the adjustment diagram. The purpose of this adjustment is to set the level of the L+R signal before being mixed with the processed L−R signal. Proceed as follows to set the L+R gain.

1. Apply an L+R signal to the MPX input.
2. Press the MAIN switch S904 and AUTO STEREO switch S905.
3. Adjust L+R control RV106 for a reading of 0.49 V p-p at pins 1 and 4 (or pins 9 and 10) of IC104.

5–4.5 L−R gain adjustment

Figure 5–11 is the adjustment diagram. The purpose of this adjustment is to set the level of the L−R signal before noise-reduction processing. This is not to be confused with the separation adjustment (Sec. 5–4.9), which sets the level after processing.
Proceed as follows to set the L−R gain.

1. Apply an L−R signal to the MPX input.
2. Press the MAIN switch S904 and AUTO STEREO switch S905.
3. Check that the STEREO lamp D901 turns on, indicating that a stereo pilot signal is present at the MPX input.
4. Adjust L−R control RV103 for a reading of 332 ±10 mV rms at NR IN test point TP28.

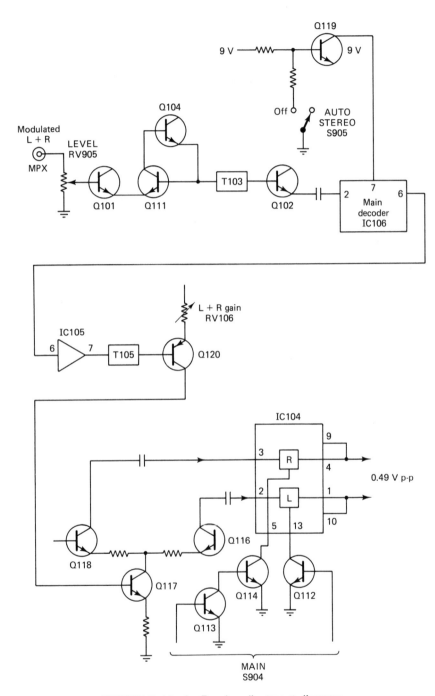

FIGURE 5-10 L+R gain adjustment diagram.

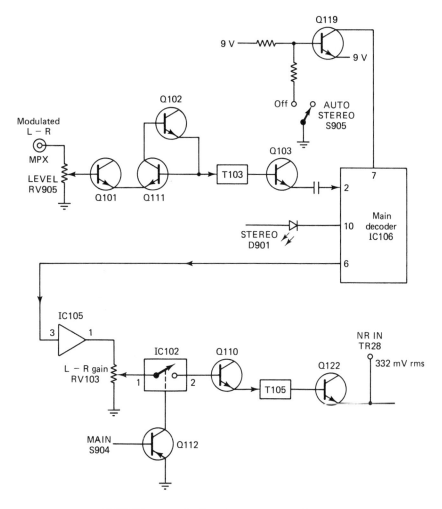

FIGURE 5–11   L–R gain adjustment diagram.

5–4.6   SAP gain adjustment

Figure 5–12 is the adjustment diagram. The purpose of this adjustment is to set the level of the SAP signal before noise-reduction processing. Note that when SAP operation is selected, the separation adjustment (Sec. 5–4.9) sets the level of the SAP signal after noise-reduction processing.

Proceed as follows to set the SAP gain.

1. Apply a SAP signal to the MPX input.

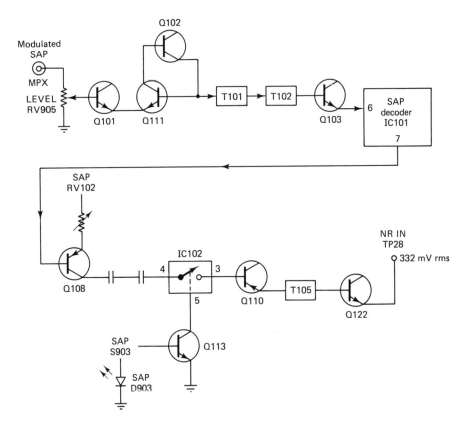

FIGURE 5-12   SAP gain adjustment diagram.

2. Press the SAP switch S903 and check that SAP lamp D903 turns on, indicating that SAP has been selected.
3. Adjust SAP control RV102 for a reading of 332 ±10 mV rms at NR IN test point TP28.

5-4.7   Noise-reduction time-constant adjustment

Figure 5-13 is the adjustment diagram. The purpose of RV501 is to set the amount of control output from dbx noise reduction IC501 for a given L−R or SAP signal amplitude. Note that IC501 is used for both L−R and SAP, depending on the operating mode selected.

Proceed as follows to set NR time constant.

1. Connect a digital multimeter across R536 as shown.
2. Adjust RV501 for a reading of 15±1 mV across R356.

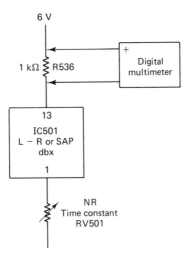

FIGURE 5–13   Noise-reduction time-constant adjustment diagram.

Note that although you are measuring voltage across R536, RV501 is set for a given current through the IC501 circuits (the more current, the more control). In this case, since R536 is 1 kΩ and you adjust for 15 mV, you are adjusting for a current of 15 mA through IC501.

5–4.8   Noise-reduction VD adjustment

Figure 5–14 is the adjustment diagram. The purpose of VD ADJ control RV502 is to set the L–R or SAP gain, after spectral processing by IC501 but before wideband processing, to produce the desired variable deemphasis or VD.

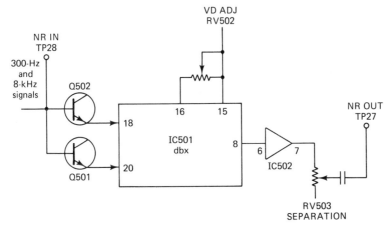

FIGURE 5–14   Noise-reduction VD adjustment diagram.

Proceed as follows to adjust the noise reduction VD.

1. Set both RV502 and SEPARATION control R503 to the midpoint or mechanical center.
2. Disconnect the MPX input, or ground the MPX input to make sure that no L−R or SAP signals are present.
3. Apply a 300-Hz sine wave to NR IN test point TP28. Set the 300-Hz sine-wave level to −24.3 dB (where 0 dB = 0.775 V).
4. Monitor the 300-Hz signal (after noise reduction) at NR OUT test point TP27. Make certain that the level at TP27 is between −35 and −27 dB. Note the actual level at TP27.
5. Change the frequency of the audio signal applied at TP28 from 300 Hz to 8 kHz. Set the amplitude of the 8-kHz signal to −17.2 dB.
6. Adjust RV502 so that the 8-kHz signal (after noise reduction) at TP27 is the actual 300-Hz value (measured in step 4) less −11.3 dB.

Keep in mind that these noise-reduction adjustments are critical to proper operation of the adapter circuits. Also, the service literature generally recommends that the VD adjustment described here, and the time-constant adjustment described in Sec. 5–4.7, be performed before the separation adjustment of Sec. 5–4.9.

5–4.9 Separation adjustment

Figure 5–15 is the adjustment diagram. The purpose of this adjustment is to set the level of the SAP or L−R signal in relation to the mono L+R signal. Since the same noise-reduction circuits are used for both SAP and L−R, it is not necessary to use either a SAP or L−R signal. Instead, the adjustment can be performed with a 300-Hz signal at the input of the noise-reduction circuits.

Proceed as follows to adjust SAP/L−R separation.

1. Apply a 300-Hz sine wave to NR IN test point TP28. Set the 300-Hz sine-wave level to −21.8 dB.
2. Press the SAP switch S903 and check that SAP lamp D903 turns on, indicating that SAP has been selected.
3. Connect an oscilloscope to pins 9 and 10 (or pins 1 and 4) of IC104.
4. Adjust SEPARATION control RV503 for a reading of 0.123 V p-p ±5 mV at both pins.

Keep in mind that this adjustment can be critical in producing good stereo sound. For a final test, measure stereo separation as described in Sec. 2–6.7.

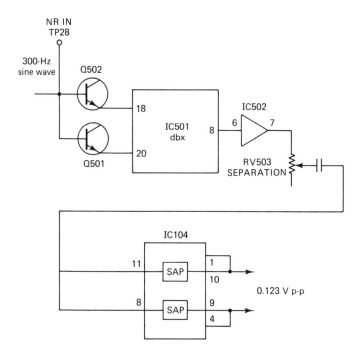

FIGURE 5–15   Separation adjustment diagram.

5–5. TROUBLESHOOTING APPROACH

All of the notes described in Secs. 1–7 and 1–8 apply to the following procedures. In each of the following trouble symptoms, it is assumed that the adapter is used with a known-good TV set and a known-good stereo system. That is, it is assumed that mono broadcasts can be played through the TV set and audio system, and that problems arise only when the adapter is used.

It is also assumed that an active TV channel has been tuned in, that both stereo and SAP signals are (supposedly) being broadcast (in addition to a mono broadcast), and that the video is good. If you do not have a good picture, the problem is probably in sections ahead of the adapter circuits (such as in the RF or IF circuits).

As a first step, check if any audio is available at the loudspeakers (either at the TV set or external stereo system), indicating a mono broadcast. Then check if the STEREO indicator is turned on, indicating a stereo broadcast.

Note that on this adapter, SAP lamp turns on when SAP is selected by the user, but does not necessarily indicate that SAP is being broadcast.

5-5.1   No audio available at the loudspeakers

If there is no audio available at the loudspeakers, but the picture is good and the STEREO indicator is on, the problem is probably in the main or L + R audio path (Fig. 5-5) or in the audio amplifiers that follow the main audio path. This can include both the audio amplifiers in the adapter (Fig. 5-7) and amplifiers in the TV set and/or external stereo. The fact that the STEREO indicator is on makes it likely (but not absolutely certain) that the sections ahead of the adapter are good.

Before you get too far into any of the circuits, make certain that the adapter controls are properly set. For example, to receive a mono L + R broadcast on both speakers, MAIN must be presssed and LINE IN must not be pressed. If LINE IN is pressed, only the audio at the rear-panel LINE IN jacks is passed, and the TV audio is cut off. If SAP is pressed and there is no SAP being broadcast, there will be no audio. If BOTH is pressed and mono is being broadcast, you should hear the mono signal on the left-channel loudspeaker.

There are several approaches that can be used at this point. However, there are two logical choices for practical troubleshooting. You can inject a composite signal with L + R at the MPX input, or inject audio at the matrix input (IC104 in Fig. 5-7). Let us start with the second approach to clear all of the audio circuits (adapter, TV set, and external stereo).

With MAIN pressed, apply audio at pins 2 and 3 of IC104. The MAIN lamp should turn on and audio should be heard on both loudspeakers. If there is no audio, inject the audio at pins 1 and 4 of IC104 and at pins 25/26 of IC107. If there is no sound on the loudspeakers with audio at pins 25/26 of IC107, suspect IC107, IC108, and the associated circuits. For example, VOLUME control RV901 could be defective (or set to full off!), as could the BALANCE, TREBLE, and BASS controls (RV904, RV903, and RV902).

If there is sound on the loudspeakers when audio is applied to pins 1 and 4 of IC104 but not when applied to pins 2 and 3, check the matrix switching circuit. This includes MAIN, SAP, and BOTH switches (S904, S903, and S902), Q112, Q113, Q114, Q115, IC104, and the associated circuits.

If the audio circuits (Fig. 5-7) appear to be good, inject a composite signal modulated with L + R at the MPX input. Then check for audio at both loudspeakers.

If the modulated signal is heard on both loudspeakers, it is reasonable to assume that the audio path through the input and main circuits (Figs. 5-4 and 5-5), and the audio amplifiers, is good. The problem is probably ahead of the adapter circuits.

If the modulated signal is not heard on the loudspeakers with an L + R signal applied, trace through the audio path using signal injection or signal tracing, whichever is most convenient. The L + R path (up to the matrix) includes LEVEL control RV905 (which could be set too low!), Q101, Q111, and Q104 shown in Fig. 5-4, and T103, Q102, IC106, IC105, T104, Q120, and Q117, shown in Fig. 5-5. Also, try correcting the problem by adjustment of RV106 as described in Sec. 5-4.4.

5-5.2   No stereo operation, mono operation is good

If there is mono audio available at the loudspeakers, and the picture is good but there is no stereo operation, check that all controls are properly set. Press MAIN switch S904 and check that MAIN lamp D904 turns on. Press AUTO STEREO switch S905 and check that STEREO lamp D901 turns on, indicating that stereo is selected and is being broadcast.

If STEREO is on but there is no stereo operation, the problem is probably in the L−R audio path shown in Fig. 5-5 or in the noise-reduction path shown in Fig. 5-6. If STEREO is off, the problem may be that no stereo is being broadcast! The quickest approach at this point is to inject a composite signal with L−R at the MPX input.

If STEREO turns on and you get stereo at the loudspeakers, the problem is probably ahead of the adapter. If STEREO does not turn on, and/or there is no stereo at the loudspeakers, the problem is in the L−R path or the noise-reduction path.

There are several points to consider when tracing through the L−R and noise-reduction paths, with either signal injection or signal tracing. If STEREO is not on, check that the base of Q119 is floating (at about 9 V). When mono is selected (AUTO STEREO off), the base of Q119 is connected to ground through R175 and S905. This causes the base of Q119 to drop, Q119 turns on, and pin 7 of IC106 goes high. This turns the VCO in IC106 off, removing the L−R carrier, and only L+R passes. With AUTO STEREO selected, the base of Q119 floats, pin 7 of IC106 drops to about 3 or 4 V, and the VCO turns on to reinsert the L−R carrier.

If STEREO is on, it is reasonable to assume that the IC106 VCO is probably functioning properly and reinserting an L−R carrier. However, this should be confirmed by further testing.

To localize the problem further, monitor for L−R audio at NR IN test point TP28 and at NR OUT test point TP27 (Fig. 5-6). If there is no audio at TP28, try correcting the problem by adjustment of VCO-2 control RV104 (Sec. 5-4.3) and L−R gain control RV103 (Sec. 5-4.5). These adjustments could cure the problem. If not, making the adjustments may lead to the fault.

Also check for L−R audio at pins 1 and 2 of IC102. If audio is available at pin 1 but not at pin 2, check that pin 13 of IC102 is high (about 9 V). If pin 13 is high but there is no audio at pin 2 of IC102, suspect IC102. If pin 13 of IC102 is not high, suspect Q112, Q114, S904, S903, S902, and the associated circuit.

If there is audio at TP28 but not at TP27, try correcting the problem by adjustment of RV501 (Sec. 5-4.7), RV502 (Sec. 5-4.8), and RV503 (Sec. 5-4.9). Then trace audio from TP28 to TP27 through Q501, Q502, IC501, and IC502.

If there is audio at TP27 but not at the loudspeakers, the problem is probably in the matrix switching circuits of Fig. 5-7. Check for processed L−R audio at pins 10 and 11 of IC102. If audio is available at pin 10 but not at pin 11, check that pin 12 of IC102 is high (about 9 V). If pin 12 is high but there is no audio at pin 11 of IC102, suspect IC102. If pin 12 of IC102 is not high, suspect Q112, Q115, S904, S903, S902, and the associated circuit.

5-5.3 No SAP operation, mono and stereo are good

If there is mono and stereo audio, and the picture is good but there is no SAP operation, check that all controls are properly set.

Press SAP switch S903 and check that SAP lamp D903 turns on. This indicates that SAP has been selected but does not necessarily show that SAP is being
broadcast. The quickest approach at this point is to inject a composite signal with
SAP at the MPX input.

If you get SAP audio at the loudspeakers, the problem is probably ahead
of the adapter (or there may not be a SAP broadcast). If there is no SAP audio
at the loudspeakers, the problem is probably in the SAP audio path of Fig. 5-4,
but could be in the matrix switching circuits of Fig. 5-7.

The circuits following the SAP audio path, between IC102 and IC104, are
probably good if you get L−R audio. You can also eliminate the input circuits of
Q101, Q104, and Q111 since these must be good to pass L−R.

Try correcting the problem by adjustment of VCO-1 control RV101 (Sec. 5-4.2)
and SAP gain control RV102 (Sec. 5-4.6). Then trace audio through T101, T102,
Q103, Q105, Q106, Q107, Q108, Q109, and IC101. Keep in mind that if there is no
SAP carrier, the SAP audio path is muted by Q105, Q106, Q107, and Q109.

Also check for SAP audio at pins 3 and 4 of IC102. If audio is available at
pin 4 but not at pin 3, check that pin 5 of IC102 is high (about 9 V). If pin 5 is
high but there is no audio at pin 3 of IC102, suspect IC102. If pin 5 of IC102 is
not high, suspect Q113, S904, S903, S902, and the associated circuits.

Finally, check for processed SAP audio at pins 8/11 and 9/10 of IC104. If
audio is available at pins 8/11 but not at pins 9/10, check that pins 6 and 12 of
IC104 are high.

If pins 6 and 12 are high but there is no audio at pins 9/10, suspect IC104.
If pin 6 is not high, suspect Q113, S904, S903, S902, and the associated circuits.
(Q113 should be off for SAP operation.) If pin 12 is not high, suspect Q112, Q115,
S904, S903, S902, and the associated circuits. (Q112 should be on and Q115 should
be off for SAP operation.)

5-5.4 No BOTH operation

If there is mono, stereo, and SAP audio, and the picture is good but it is not
possible to receive both mono and SAP simultaneously (mono on the left and SAP
on the right), and BOTH switch S902 has been pressed, the problem is localized to
the matrix switching circuits of Fig. 5-7. This is based on the assumption that both
mono and SAP are being broadcast simultaneously, or that you have injected both
mono and SAP at the MPX input.

6

SONY MTS
TV STEREO

This chapter is devoted to the decoder circuits used in typical Sony stereo-ready TV sets. Note that the circuits are referred to as MTS, or multichannel TV sound, in Sony literature. The PC board containing the stereo-TV circuits is labeled the X-board.

The decoder circuits receive their input in the form of composite audio from a sound i-f (SIF) detector or discriminator. The broadcast r-f signal is converted to an i-f signal by a tuner located on an A-board. The i-f signal is converted to a 4.5-MHz sound carrier by an i-f IC, also located on the A-board.

The composite audio is extracted from the 4.5-MHz carrier by the SIF detector IC located on the A-board. The composite audio is amplified on the A-board and applied to the input of the decoder circuits on the X-board.

The output of the decoder circuits is applied to dual speakers (mounted on each side of the TV set) through a stereo amplifier and audio controller (also located on the X-board).

6-1. OPERATING (USER) CONTROLS AND INDICATORS

Two front-panel operating (or user) controls and two front-panel indicators are associated with the decoder circuits. Figure 6–1 shows a typical arrangement for the controls and indicators. The following paragraphs describe the control and indicator functions.

Press the AUTO STEREO button to select automatic stereo operation. Under these conditions, the STEREO lamp should turn on when stereo is being broadcast.

Press the MTS button to select stereo, SAP, or both modes of operation. The choice of operating mode is shown by an on-screen display (upper right-hand side of the TV screen). Every push of the MTS button changes the on-screen display, as shown in Fig. 6–1 (MAIN to SAP to BOTH and back to MAIN, in that order).

The on-screen display is produced by a character generator, which, in turn, is controlled by microprocessors. The microprocessors receive commands from the MTS button and produce corresponding signals to the character generator.

The same microprocessors also receive commands from the VOLUME, BALANCE, TREBLE, and BASS operating controls, and apply corresponding signals to the audio controller on the X-board. We do not cover these controls on audio circuits in this book, since the circuits are not a direct part of the MTS system.

6-1.1 Typical operating sequence

To receive a stereo program, press AUTO STEREO. In this position, the STEREO lamp turns on when a stereo broadcast is received.

To hear the program in stereo, press the MTS button until MAIN appears at the on-screen display. The stereo sound should be heard from the built-in left and right speakers.

Note that there may be cases of stereo broadcasts where excessive noise is heard, usually due to a weak incoming signal. Often, such noise can be eliminated by switching the AUTO STEREO to off and returning to a monaural mode (from both speakers).

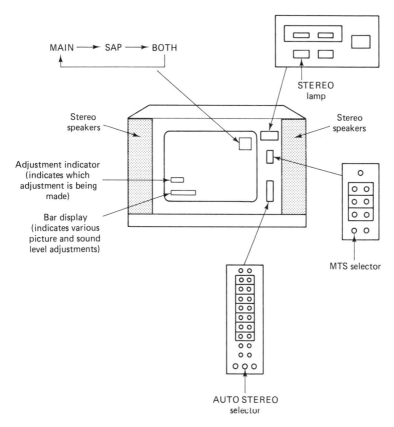

FIGURE 6-1   Typical arrangement for controls and indicators.

To hear a SAP program, press the MTS button until SAP appears at the on-screen display. The SAP sound should be heard from both built-in speakers.

To hear both SAP and main-program material simultaneously, press the MTS button until BOTH appears at the on-screen display.

In the BOTH mode, main-program material should be heard from the left speaker, while SAP is heard from the right speaker. The main-program material is in monaural form.

Note that if SAP is not being broadcast, there will be no sound from the right speaker. If there is no SAP carrier, the SAP audio channel is muted (to prevent background noise or any other audio from passing to the right speaker).

6-2. CIRCUIT DESCRIPTIONS

All of the notes described in Sec. 1-7 apply to the following descriptions. Note that the circuits covered here are similar to those of the Sony MTS adapter described in Chapter 5. However, there are subtle differences that must be considered in troubleshooting.

6–2.1 Composite input and SAP audio path

Figure 6–2 shows the composite input and SAP audio path in simplified form. The composite input from the SIF detector is applied to both the SAP and main-program audio paths through connector X5. The composite signal is amplified by Q1101 and applied to the SAP audio path through filters T1101 and T1102.

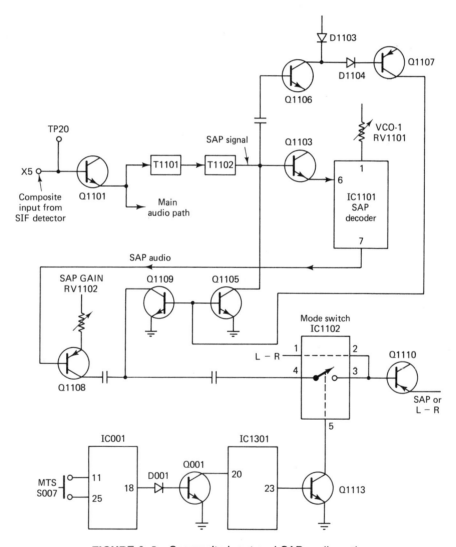

FIGURE 6–2 Composite input and SAP audio path.

The SAP signal (carrier plus audio) is applied to SAP decoder IC1101 through amplifier Q1103. SAP audio is taken from pin 7 of IC1101 and applied to mode switch IC1102 through amplifier Q1108.

VCO-1 control RV101 sets the frequency of the VCO within IC1101 (78.67 kHz). Gain or level control RV1102 sets the gain or level of SAP amplifier Q1108.

A SAP-mute circuit consisting of Q1105, Q1106, Q1107, and Q1109 disables the SAP audio path when there is no SAP carrier signal.

The SAP signal is amplified by Q1106, rectified by D1103/D1104, and applied to Q1107. If SAP is present, the base of Q1107 is high (about 13.5 V) and the collector is zero. This keeps Q1105/D1109 off.

If SAP is not being broadcast, the base of Q1107 goes low (about 8 V), and the collector rises to about 9 V. This drives both Q1105 and Q1109 into conduction, bypassing any signals in the SAP path to ground.

The SAP output from mode switch IC1102 at pin 3 is applied to amplifier Q1110, along with the L−R signal (Sec. 6-2.2). The SAP signal at pin 3 of IC1102 is controlled by Q1113, which, in turn, receives commands from controller IC1301. The MTS button or switch S007 connects pins 11 and 25 of key interface IC001 when pressed. This applies commands to pin 20 of controller IC1301. The commands switch pins 22 and 23 of IC1301 high and low as required to select the desired operating mode (MAIN, SAP, or BOTH).

When either SAP or BOTH is selected, the base of Q1113 goes low (zero) and the collector rises to 9 V, connecting pins 3 and 4 of IC1102. When SAP is off, the base of Q1113 goes high (about 0.7 V) and the collector drops to zero, disconnecting pins 3 and 4 of IC1102.

6-2.2 Main L+R and L−R audio paths

Figure 6-3 shows the L+R and L−R audio paths in simplified form. The L+R and L−R signals are passed by filter T1103 and applied to stereo decoder IC1106 through buffer Q1124. The signals at pins 5 and 6 of IC1106 are applied to amplifiers within IC1105 to become the L+R and L−R audio.

L+R audio is applied to the mixer circuit (Sec. 6-2.4) through IC1105, T1104, and Q1120. MAIN control RV1106 sets the gain or level of the L−R audio. VCO-2 control RV1104 sets the frequency of the L−R VCO within IC1106.

The L−R output from mode switch IC1102 at pin 2 is applied to amplifier Q1110, along with the SAP signal (Sec. 6-2.1). The L−R signal at pin 2 of IC1102 is controlled by Q1112, which, in turn, is controlled by IC1301, IC001, and MTS switch S007.

When MAIN is selected, the base of Q1112 goes low (zero) and the collector rises to 8.5 V, connecting pins 1 and 2 of IC1102. When MAIN is off, the base of Q1112 goes high (about 0.6 V) and the collector drops to zero, disconnecting pins 1 and 2 of IC1102.

The choice between monaural (L+R) or stereo (L−R) is determined by the

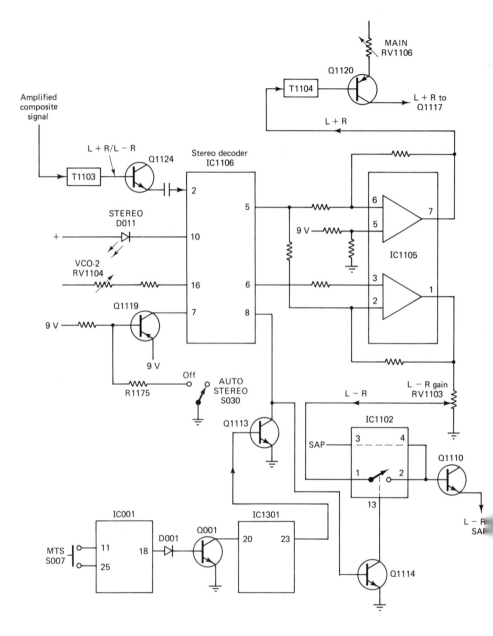

FIGURE 6-3 Main L+R and L−R audio paths.

Q1119 signal applied at pin 7 of IC1106. When AUTO STEREO switch S030 is set to mono, the base of Q1119 goes low. This drives Q1119 into conduction and applies a high to IC1106 at pin 7. Under these conditions, the L−R VCO is disabled, only L+R passes, and pin 10 of IC1106 goes high, turning off STEREO lamp D011.

When AUTO STEREO switch S030 is set to auto, the base of Q1119 is floating (about 9 V) and Q1119 remains off, dropping pin 7 of IC1106 to about 3.8 V. Under these conditions, the L−R VCO is turned on, both L+R and L−R pass, and pin 10 of IC1106 goes low (zero), turning on STEREO lamp D011.

Note that pin 10 of IC1106 does not go low unless there is a pilot signal present. So D011 may not turn on even though AUTO STEREO has been selected.

When SAP or BOTH are selected, pin 23 of IC1301 goes high, Q1113 turns on, and pin 8 of IC1106 goes low, disabling pin 6. Under these conditions, only L+R is available at IC1105.

6-2.3 dbx noise-reduction path

Figure 6-4 shows the dbx noise-reduction path in simplified form. Note that both SAP and L−R use the same noise-reduction path.

The SAP/L−R signals are applied to dbx noise reduction IC1501 through Q1110, T1105, Q1122, Q1501, and Q1502.

SAP/L−R audio from the emitter of Q1502 is applied to the spectral VCA at pin 18 of IC501 through a fixed deemphasis network C1503/C1504/R1507/R1508). Audio signals from the emitter of Q1122 (in the range from 100 Hz to 3 kHz) are applied to the wideband rms detector at pin 3 of IC1501. C1505/C1506 and R1504/R1505 form the wideband BPF. Audio from the emitter of Q1501 (in the range 4 to 15 kHz) is applied to the spectral rms detector at pin 20 if IC1501. C1509/C1510 and R1512/R1513/R1514 form the spectral BPF.

Two rms detectors in IC1501 receive power from a constant-current generator. The output of this generator is adjustable by T CONT control RV1501, the timing or time-constant control, connected at pin 1 of IC1501. The setting of RV1501 determines the amount of control output from the rms detectors for a given signal amplitude.

Spectral processing is performed by the spectral VCA at pin 18 of IC1501, controlled by the output of the spectral rms detector. The signal is then applied to an amplifier within IC1502 through C1513, an op-amp within IC1501 and C1515. VD ADJ control RV1502 sets signal level after spectral processing, but before wideband processing, by setting the gain of the op-amp.

The amplified signal is then applied to the wideband VCA at pin 5 of IC1501. Wideband processing is performed by the wideband VCA, controlled by the output of the wideband rms detector.

The processed signal is then applied to another op-amp within IC1501 through C1522. The op-amp output at pin 8 of IC1501 is applied to the mixer and audio output circuit (Sec. 6-2.4) through another amplifier in IC1502. Separation control RV1503 sets the gain or level of the processed signal (and thus the separation) when

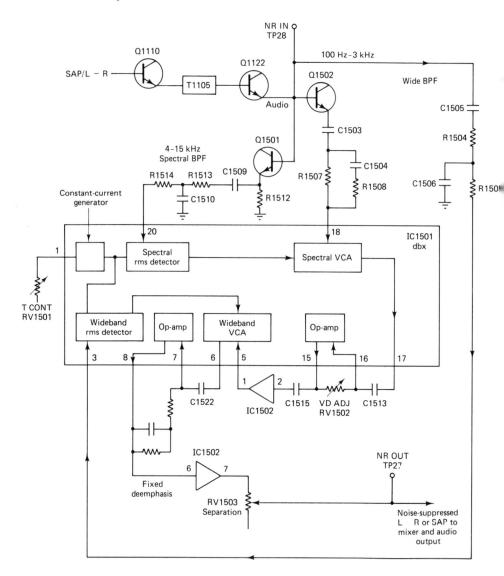

FIGURE 6–4 dbx noise-reduction path.

L – R is being processed. The network at pins 7 and 8 of IC1501 produce the required fixed deemphasis.

6–2.4 Mixer and audio output path

Figure 6–5 shows the mixer and audio output path in simplified form. The L + R, L – R, and SAP signals are applied to audio control IC1107 through mode switch

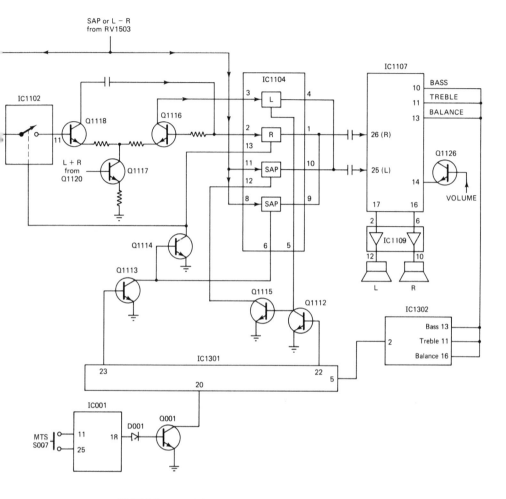

FIGURE 6-5   Mixer and audio output path.

IC1102, mode switch IC1104, and the associated mixer circuits (Q1116, Q1117, and Q1118).

The L+R signal from the emitter of Q1120 is applied to the base of Q1117. The SAP/L−R signal from RV1503 is applied to pins 8 and 11 of IC1104 and to pin 10 of IC1102.

When MAIN is selected with the MTS button and there is no stereo and/or when AUTO STEREO is off, only the L+R signal from Q1117 is applied to pins 25 and 26 of IC1107 through Q1116, Q1118, and IC1104 (pins 1/2/3/4). This produces a monaural input to IC1107 and mono output to the speakers through audio amplifier IC1109.

When MAIN is selected with the MTS button, there is stereo being broad-cast, and AUTO STEREO is on, L−R from RV1503 is combined with L+R from Q1117 to produce a stereo input to IC1107 (pin 25 left, pin 26 right). This results in a stereo output to the loudspeakers.

When SAP is selected with the MTS button, SAP (if any) from RV1503 is applied to IC1107 through IC1104 (pins 10/11 and 8/9) to produce a monaural SAP output to the loudspeakers. L+R signals are disconnected from IC1107 since pins 1/2 and 3/4 of IC1104 are open.

When BOTH is selected, pin 6 of IC1104 goes high. The SAP signal from RV1503 is applied through pins 8 and 9 of IC1104 to pin 26 of IC1107, producing a SAP output on the right channel.

When BOTH is selected, pin 5 of IC1104 goes high. The L+R signal from Q1117 is applied through Q1116 and pins 3/4 of IC1104 to pin 25 of IC1107, producing an L+R output on the left channel.

SAP cannot pass through pins 10 and 11 of IC1102 since pin 12 of IC1102 is low. (When Q113 turns off and pin 6 of IC1104 goes high, Q114 turns on, causing pin 12 of IC1102 to go low and opening pins 10 and 11 of IC1102.)

6-3. TYPICAL TEST/ADJUSTMENT PROCEDURES

All of the notes described in Sec. 1–7.6 apply to the following test/adjustment procedures.

6-3.1 SAP decoder VCO-1 adjustment

Figure 6–6 is the adjustment diagram. The purpose of the SAP decoder IC1101 is to produce audio at pin 7 when there is a SAP carrier and audio at input pin 6. IC1101 contains a VCO operating at 78.67 kHz (the SAP carrier frequency) that must be set when free-running. The VCO can be set by VCO-1 control RV1101 and can be monitored at pin 2 of IC1101.

Proceed as follows to adjust the SAP decoder.

1. Short circuit pins 5 and 6 of IC1101 with a jumper. This eliminates any possibility of a SAP carrier or pilot entering the circuits. If the carrier or pilot

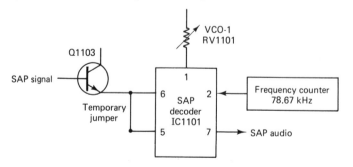

FIGURE 6-6   SAP decoder VCO-1 adjustment diagram.

is present, it is possible that the VCO will lock on to the incoming signal, even though the VCO is not exactly on-frequency when free-running.

2. Connect a frequency counter to pin 2 of IC1101.

3. Adjust RV1101 for a reading of 78.67 kHz ± 300 Hz.

4. Remove the jumper. Check that the reading at pin 2 of IC1101 remains at 78.67 kHz.

6–3.2 Stereo decoder VCO-2 adjustment

Figure 6–7 is the adjustment diagram. One purpose of stereo decoder IC1106 is to reinsert a carrier into the AM stereo L–R sidebands and produce corresponding audio output at pins 5 and 6. The missing L–R carrier is at 31.468 and is produced by a VCO within IC1106. Since there is no L–R carrier available to IC1106, the VCO is usually locked to the 15.734-kHz pilot (or a multiple).

In the circuit of Fig. 6–7, the VCO is set by VCO-2 control RV1104 at pin 16 of IC1106 and can be monitored at the slider of RV1104. The VCO signal at RV1104 is 62.936 kHz, even though the L–R requires a carrier of 31.468 kHz.

Proceed as follows to adjust the stereo decoder.

1. Disconnect the X5 connector. This removes all signals to IC1106, including the 15.734-kHz pilot. If the pilot (or multiple) is present, it is possible that

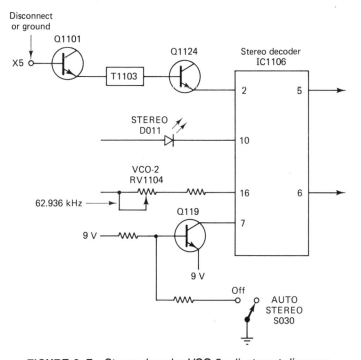

FIGURE 6–7 Stereo decoder VCO-2 adjustment diagram.

the VCO will lock on to the incoming signal, even though the VCO is not exactly on-frequency when free-running.

2. Press the AUTO STEREO switch S030. This enables the VCO in IC1106. However, STEREO lamp D011 should not turn on since there is no pilot.

3. Connect a frequency counter to the slider of RV1104.

4. Adjust RV1104 for a reading of 62.926 kHz ± 300 Hz.

5. Replace the X5 connector. Check that the reading at RV1104 remains at 62.936 kHz.

6–3.3 L+R gain adjustment

Figure 6–8 is the adjustment diagram. The purpose of this adjustment is to set the level of the L+R signal before being mixed with the processed L−R signal. Proceed as follows to set the L+R gain.

1. Apply an L+R signal to the X5 input.

2. Press the MTS switch S007 until MAIN appears at the on-screen display. Press the AUTO STEREO switch S030.

3. Adjust MAIN control RV1106 for a reading of 0.6 V p-p at pins 1 and 4 (or pins 9 and 10) of IC1104.

6–3.4 L−R gain adjustment

Figure 6–9 is the adjustment diagram. The purpose of this adjustment is to set the level of the L−R signal before noise-reduction processing. This is not to be confused with the separation adjustment (Sec. 6–3.8), which sets the level after processing. Proceed as follows to set the L−R gain.

1. Apply an L−R signal to the X5 input.

2. Press the MTS switch S007 until MAIN appears at the on-screen display. Press the AUTO STEREO switch S030.

3. Check that the STEREO lamp D011 turns on, indicating that a stereo pilot signal is present at the X5 input.

4. Adjust L−R control RV1103 for a reading of 332 ± 10 mV rms at NR IN test point TP28.

6–3.5 SAP gain adjustment

Figure 6–10 is the adjustment diagram. The purpose of this adjustment is to set the level of the SAP signal before noise-reduction processing. Note that when SAP operation is selected, the separation adjustment (Sec. 6–3.8) sets the level of the SAP signal after noise-reduction processing.

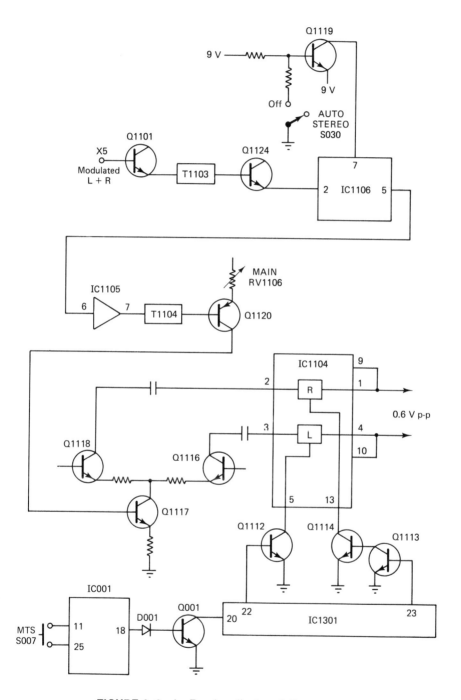

FIGURE 6-8 L+R gain adjustment diagram.

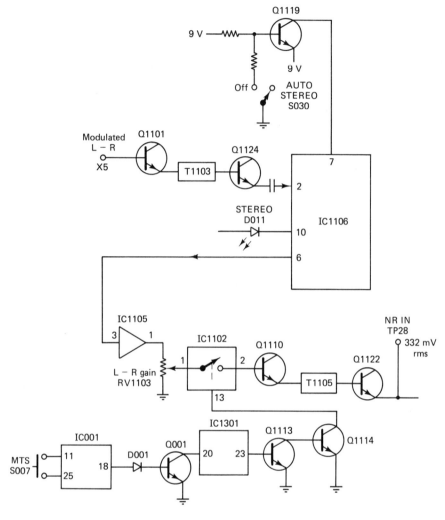

FIGURE 6-9 L–R gain adjustment diagram.

Proceed as follows to set the SAP gain.

1. Apply a SAP signal to the X5 input.
2. Press the MTS switch S007 until SAP appears at the on-screen display.
3. Adjust SAP GAIN RV1102 for a reading of 332 ± 10 mV rms at NR IN test point TP28.

6–3.6 *Noise-reduction time-constant adjustment*

Figure 6–11 is the adjustment diagram. The purpose of RV1501 is to set the amount of control output from dbx noise reduction IC1501 for a given L–R or SAP signal

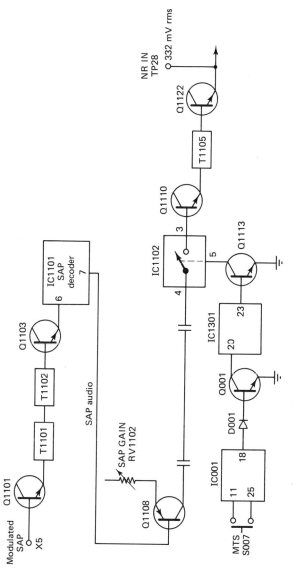

FIGURE 6-10 SAP gain adjustment diagram.

145

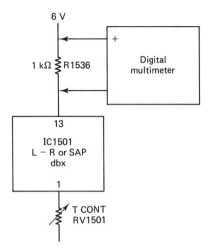

FIGURE 6–11 Noise-reduction VD adjustment diagram.

amplitude. Note that IC1501 is used for both L−R and SAP, depending on the operating mode selected.

Proceed as follows to set NR time constant.

1. Connect a digital multimenter across R1536 as shown.
2. Adjust T CONT RV1501 for a reading of 15 ± 1 mV across R1536.

Note that although you are measuring voltage across R1536, RV1501 is set for a given current through the IC1501 circuits (the more current, the more control). In this case, since R1536 is 1 kΩ and you adjust for 15 mV, you are adjusting for a current of 15 mA through IC1501.

6–3.7 Noise-reduction VD adjustment

Figure 6–12 is the adjustment diagram. The purpose of VD ADJ RV1502 is to set the L−R or SAP gain, after spectral processing by IC1501 but before wideband processing, to produce the desired variable deemphasis or VD.

Proceed as follows to adjust the noise-reduction VD.

1. Set both RV1502 and SEPARATION control RV1503 to the midpoint or mechanical center.
2. Cut off Q1122. This can be done by connecting the emitter of Q1122 (or TP28) to the 9-V line through a 1-kΩ resistor. This will make sure that no L−R or SAP signals are present at IC1501.
3. Apply a 300-Hz sine wave to NR IN test point TP28. Set the 300-Hz sine-wave level to −24.3 dB (where 0 dB = 0.775 V).

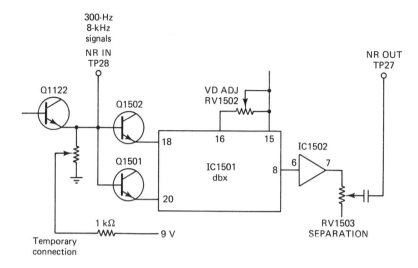

FIGURE 6–12   Noise-reduction VD adjustment diagram.

4. Monitor the 300-Hz signal (after noise reduction) at NR OUT test point TP27. Make certain that the level at TP27 is between −35 and −27 dB. Note the actual level at TP27.

5. Change the frequency of the audio signal applied at TP28 from 300 Hz to 8 kHz. Set the amplitude of the 8-kHz signal to −17.2 dB.

6. Adjust RV1502 so that the 8-kHz signal (after noise reduction) at TP27 is the actual 300-kHz value (measured in step 4) less −11.3 dB.

Keep in mind that these noise-reduction adjustments are critical to proper operation of the MTS circuits. Also, the service literature generally recommends that the VD adjustment described here, and the time-constant adjustment described in Sec. 6–3.6, be performed before the separation adjustment of Sec. 6–3.8.

6–3.8 Separation adjustment

Figure 6–13 is the adjustment diagram. The purpose of this adjustment is to set the level of the SAP or L−R signal in relation to the mono L+R signal. Since the same noise-reduction circuits are used for both SAP and L−R, it is not necessary to use either a SAP or L−R signal. Instead, the adjustment can be performed with a 300-Hz signal at the input of the noise-reduction circuits.

Proceed as follows to adjust SAP/L−R separation.

1. Apply a 300-Hz sine wave to NR IN test point TP28. Set the 300-Hz sine-wave level to −21.8 dB.

2. Press the MTS switch S007 until SAP appears at the on-screen display.

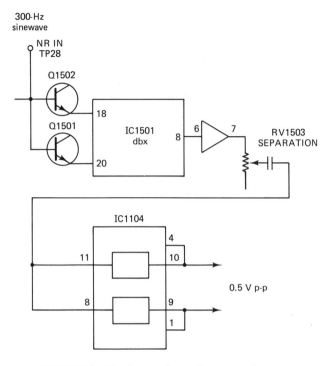

FIGURE 6-13   Separation adjustment diagram.

3. Connect an oscilloscope to pins 9 and 10 (or pins 1 and 4) of IC1104.

4. Adjust SEPARATION control RV1503 for a reading of 0.5 V p-p ± 5 mV p-p at both pins.

Keep in mind that this adjustment can be critical in producing good stereo sound. For a final test, measure stereo separation as described in Sec. 2-6.7.

6-4. TROUBLESHOOTING APPROACH

All of the notes described in Secs. 1-7 and 1-8 apply to the following procedures. In each of the following trouble symptoms, it is assumed that an active TV channel has been tuned in, that both stereo and SAP signals are (supposedly) being broadcast (in addition to a mono broadcast), and that the video is good. If you do not have a good picture, the problem is probably in sections ahead of the MTS circuits (such as in the RF or IF circuits).

As a first step, check if any audio is available at the loudspeakers (indicating a mono broadcast). Then check if the STEREO lamp is turned on (indicating a stereo broadcast signal).

Note that on this TV set, SAP appears at the on-screen display when selected by the user (with the MTS button). However, this does not necessarily indicate that SAP is being broadcast.

6–4.1 No audio available at the loudspeakers

If there is no audio available at the loudspeakers, but the picture is good and the STEREO lamp is turned on, the problem is probably in the main or L + R audio path, or in the audio circuits that follow the main audio path. This can include the audio controller IC1107, audio amplifier IC1109, D/A converter IC1302, and controller IC1301 (as well as the front-panel control buttons or keys and the key interface IC001). The fact that the STEREO lamp is turned on makes it likely (but not absolutely certain) that the sections ahead of the MTS circuits are good.

Before you get too far into any of the circuits, make certain that the TV set controls are properly set. For example, to receive a mono L + R broadcast on both speakers, the MTS button must be pressed until MAIN appears at the on-screen display, and AUTO STEREO must not be pressed.

If AUTO STEREO is pressed (with MAIN at the on-screen display), you should get stereo at the loudspeakers (if stereo is being broadcast and the STEREO lamp is on). If SAP appears at the on-screen display and there is no SAP being broadcast, there should be no audio. If BOTH appears at the on-screen display and mono is being broadcast, you should hear the mono signal on the left-channel loudspeaker.

There are several approaches that can be used at this point. However, there are two logical choices for practical troubleshooting. You can inject a composite signal with L + R at the MTS circuit input (at connector X5), or inject audio at the matrix input (IC1104 in Fig. 6–5). Let us start with the second approach to clear all of the audio circuits (controller, amplifiers, loudspeakers, etc.).

Press the MTS button until MAIN appears at the on-screen display. Apply audio at pins 2 and 3 of IC1104. Audio should be heard on both loudspeakers.

If there is no audio, inject audio at pins 1 and 4 of IC1104 and at pins 25/26 of IC1107. If there is no sound on the loudspeakers with audio at pins 25/26 of IC1107, suspect IC1107, IC1109, and the associated circuits.

For example, the front-panel VOLUME could be defective (or set to full off!). The problem could also be between VOLUME and IC1107 (including Q1126).

If there is sound on the loudspeakers when audio is applied to pins 1 and 4 of IC1104 but not when applied to pins 2 and 3, check the matrix switching circuit. This includes the circuits between MTS switch S007 (through IC001, D001, Q001, IC1301, Q1112, Q1113, Q1114, Q1115) and the switch IC1104.

If the audio circuits (Fig. 6–5) appear to be good, inject a composite signal modulated with L + R at the X5 input. Then check for audio at both loudspeakers.

If the modulated signal is heard on both loudspeakers, it is reasonable to assume that the audio path through the input and main circuits (Figs. 6–2 and 6–3), and the audio amplifiers, is good. The problem is probably ahead of the MTS circuits.

If the modulated signal is not heard on the loudspeakers with an L + R signal applied, trace through the audio path using signal injection or signal tracing, whichever is most convenient. The L + R path (up to the matrix) includes Q1101, T1103, Q1124, IC1106, IC1105, T1104, Q1120, and Q1117 (Figs. 6-2 and 6-3). Also, try correcting the problem by adjustment of RV1106 as described in Sec. 6-3.3.

6-4.2 No stereo operation, mono operation is good

If there is mono audio available at the loudspeakers, and the picture is good but there is no stereo operation, check that all controls are properly set.

Press MTS switch S007 until MAIN appears at the on-screen display. Press AUTO STEREO switch S030 and check that STEREO lamp D011 turns on, indicating that stereo is selected and is being broadcast.

If STEREO is on but there is no stereo operation, the problem is probably in the L − R audio path shown in Fig. 6-3, or in the noise-reduction path shown in Fig. 6-4. If STEREO is off, the problem may be that no stereo is being broadcast! The quickest approach at this point is to inject a composite signal with L − R at the X5 input.

If STEREO turns on, and you get stereo at the loudspeakers, the problem is probably ahead of the MTS circuits. If STEREO does not turn on, and/or there is no stereo at the loudspeakers, the problem is in the L − R path or the noise-reduction path.

There are several points to consider when tracing through the L − R and noise-reduction paths, with either signal injection or signal tracing. If STEREO is not on, check that the base of Q1119 is floating (at about 9 V). When mono is selected (AUTO STEREO off), the base of Q1119 is connected to ground through R1175 and AUTO STEREO switch S030. This causes the base of Q1119 to drop, Q1119 turns on, and pin 7 of IC1106 goes high. This turns the VCO in IC1106 off, removing the L − R carrier, and only L + R passes. With AUTO STEREO selected, the base of Q1119 floats, pin 7 of IC1106 drops to about 3 or 4 V, and the VCO turns on to re-insert the L − R carrier.

If STEREO is on, it is reasonable to assume that the IC1106 VCO is probably functioning properly and reinserting a L − R carrier. However, this should be confirmed by further testing.

To localize the problem further, monitor for L − R audio at NR IN test point TP28 and at NR OUT test point TP27 (Fig. 6-4). If there is no audio at TP28, try correcting the problem by adjustment of VCO-2 control RV1104 (Sec. 6-3.2) and L − R gain control RV1103 (Sec. 6-3.4). These adjustments could cure the problem. If not, making the adjustments may lead to the fault.

Also check for L − R audio at pins 1 and 2 of IC1102. If audio is available at pin 1 but not at pin 2, check that pin 13 of IC1102 is high (about 9 V). If pin 13 is high but there is no audio at pin 2 of IC1102, suspect IC1102. If pin 13 of IC1102 is not high, suspect Q1113, Q1114, and IC1301.

Pin 23 of IC1301 should be high when MAIN is selected (if not, suspect the MTS switch S007, IC001, Q001, and D001). With pin 23 of IC1301 high, Q1113 is turned on, Q1114 is turned off, and pin 13 of IC1102 goes high to connect pins 1 and 2.

If there is audio at TP28 but not at TP27, try correcting the problem by adjustment of RV1501 (Sec. 6–3.6), RV1502 (Sec. 6–3.7), and RV1503 (Sec. 6–3.9). Then trace audio from TP28 to TP27 through Q1501, Q1502, IC1501, and IC1502. If there is audio at TP27 but not at the loudspeakers, the problem is probably in the mixer and audio output circuits of Fig. 6–5.

Check for processed L−R audio at pins 10 and 11 of IC1102. If audio is available at pin 10 but not at pin 11, check that pin 12 of IC1102 is high (about 9 V). If pin 12 is high but there is no audio at pin 11 of IC1102, suspect IC1102. If pin 12 of IC1102 is not high, suspect Q1113, Q1114, and IC1301.

6–4.3 No SAP operation, mono and stereo are good

If there is mono and stereo audio, and the picture is good but there is no SAP operation, check that all controls are properly set.

Press the MTS until SAP appears at the on-screen display. This indicates that SAP has been selected, but does not necessarily show that SAP is being broadcast.

The quickest approach at this point is to inject a composite signal with SAP at the X5 input. If you get SAP audio at the loudspeakers, the problem is probably ahead of the MTS circuits (or there may not be a SAP broadcast).

If there is no SAP audio at the loudspeakers, the problem is probably in the SAP audio path of Fig. 6–2, but could be in the mixer and audio output circuits of Fig. 6–5.

The circuits following the SAP audio path between IC1102 and IC1104, are probably good if you get L−R audio. You can also eliminate the input circuit of Q1101 since this must be good to pass L−R.

Try correcting the problem by adjustment of VCO-1 control RV1101 (Sec. 6–3.1) and SAP gain control RV1102 (Sec. 6–3.5). Then trace audio through T1101, T1102, Q1103, Q1105, Q1106, Q1107, Q1108, Q1109, and IC1101. Keep in mind that if there is no SAP carrier, the SAP audio path is muted by Q1105, Q1106, Q1107, and Q1109.

Also check for SAP audio at pins 3 and 4 of IC1102. If audio is available in pin 4 but not at pin 3, check that pin 5 of IC1102 is high (about 9 V). If pin 5 is high but there is no audio at pin 3 of IC1102, suspect IC1102. If pin 5 of IC1102 is not high, suspect Q1113 and IC1301.

Pin 23 of IC1301 should be low when SAP is selected (if not, suspect the MTS switch S007, IC001, Q001, and D001). With pin 23 of IC1301 low, Q1113 is turned off and pin 5 of IC1102 goes high to connect pins 3 and 4.

Finally, check for processed SAP audio at pins 8/11 and 9/11 of IC1104. If audio is available at pins 8/11 but not at pins 9/10, check that pins 6 and 12 of IC1104 are high.

If pins 6 and 12 are high but there is no audio at pins 9/10, suspect IC1104. If pins 6 and 12 are not high, suspect Q1112/Q1115, IC1301, and the associated circuits.

6–4.4   No BOTH operation

If there is mono, stereo, and SAP audio, and the picture is good but it is not possible to receive both mono and SAP simultaneously (mono on the left, and SAP on the right), and BOTH has been selected by the MTS switch, the problem is localized to the matrix switching circuits of Fig. 6–5. This is based on the assumption that both mono and SAP are being broadcast simultaneously, or that you have injected both mono and SAP at the X5 input.

6–5.   SOME ADDITIONAL CIRCUITS

The following paragraphs describe some additional MTS circuits found in Sony television. These circuits are quite different from those described thus far.

6–5.1   MPX decoder

Figure 6–14 shows the MPX decoder circuits used in some Sony TV (such as the KV-1981R). Note that the majority of the circuits are contained within a single IC, designated as the CX20112.

The functions performed by the CX20112 IC include the SAP bandpass filter (5fH), the SAP VCO (5fH0), and the SAP FM detector. Also included are the stereo low-pass filter (3fH down), the pilot detector (1fH), the VCO (4fH) and the necessary dividers, and the AM stereo decoder. All necessary switching for directing SAP or stereo to the noise-reduction circuit, and for directing the appropriate signal to the built-in matrix circuit, is also included.

Operating modes are selected by the switch control inputs at pins 16, 17, and 20 of the IC. The possible operating modes are: auto stereo or mono, main, SAP, or both.

A high at pin 20 of the IC forces the matrix into mono operation.

A low at pin 17 of the IC switches the matrix to stereo if the stereo pilot is present. If there is no stereo pilot, the matrix operates in mono mode, passing the main audio channel (L+R).

A low at pin 16 of the IC switches the matrix to process the SAP audio channel in mono mode (SAP at both loudspeakers).

A low at both pins 16 and 17 of the IC causes the matrix to process SAP on the right channel, and main L+R audio on the left channel.

The SAP mute circuit Q872/Q877 operates when SAP is present to enable the SAP detector by making pin 14 of the IC low.

The signal path for the L+R audio is in at pin 1, out at pin 37, in at pin 34, through an internal L+R amplifier and low-pass filter, out at pin 29, through the 75-μs deemphasis circuit, and in at pin 28 to the internal matrix.

The signal path for stereo L−R is in at pin 1, out at pin 37, in at pin 34, to the stereo detector, out at pin 40, through the level set (RV816), and in at pin 41.

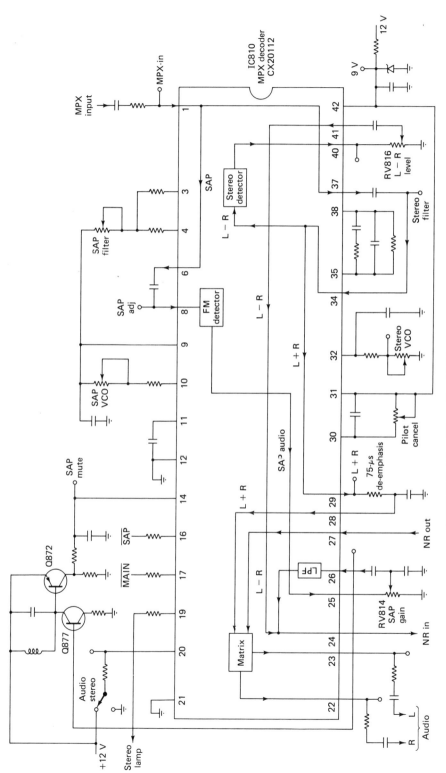

FIGURE 6-14 Alternative MPX decoder circuits.

The L–R then passes through a low-pass filter, out at pin 24 to the noise-reduction circuit, back from the noise-reduction circuit, and in at pin 27 to the matrix. The matrixed L + R and L – R signals develop the left and right audio outputs at pins 22 and 23.

The SAP signal flow is in at pin 1 to the SAP bandpass filter, out at pin 6, in at pin 8 to the FM detector. The detected SAP is then output at pin 25, across the level set (RV814), in at pin 26, through another internal low-pass filter, out to the noise-reduction circuit at pin 24, and back from noise reduction at pin 27 to the matrix where SAP is processed as mono audio (both speakers).

In the BOTH mode, the main (L + R) and SAP signal flows are the same, except that the matrix processes SAP as right-channel audio and L + R as left-channel audio.

6–5.2 Noise reduction

Figure 6–15 shows the noise-reduction circuits used in some Sony TVs (such as the KV-1981R). Note that the majority of the circuits are contained within two ICs. These ICs are similar to the noise-reduction and matrix ICs described thus far in this chapter and are common to those used in the TV sets of many manufacturers, so we will not describe the circuits in detail. However, we include the circuits here to show tie-in to the decoder circuits described in Sec. 6–5.2 (Fig. 6–14).

The SAP and stereo signal (L–R) are input to the noise-reduction circuit at Q831. Transistor Q831 and the associated circuit provide the fixed deemphasis.

The SAP and L–R are also applied through a low-pass filter to pin 3 of the noise-reduction IC (to one rms detector), and through Q830 and a high-pass filter to pin 20 of the noise-reduction–IC (to the other rms detector).

Spectral deemphasis is performed on the signal by the noise-reduction IC. The output, after spectral deemphasis, is applied to pin 2 of the matrix IC from pin 15 of the noise-reduction IC. The output from pin 1 of the matrix IC is applied to the wideband processing circuits of the noise-reduction IC at pin 5.

Wideband processing is performed on the signal by the noise-reduction IC. The output, after wideband processing, is applied to pin 6 of the matrix IC from pin 8 of the noise-reduction IC. The output from pin 7 of the matrix IC is applied to the decoder circuits.

The time-constant control RV830 is connected at pin 1 of the noise-reduction IC and sets the level of current produced by the constant-current source or generator.

The VD adjust control RV831 is connected across pins 15 and 16 of the noise-reduction IC and sets the level of the spectral deemphasis op-amp.

The separation control RV832 is connected across pins 6 and 7 of the matrix IC and sets the gain of the noise-processed signal.

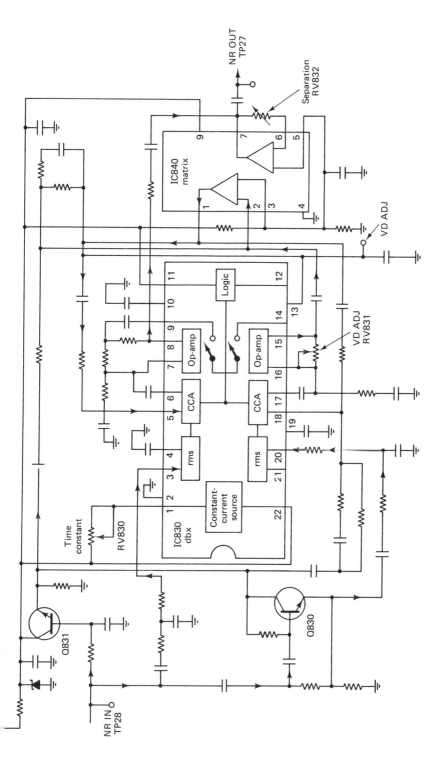

FIGURE 6-15 Alternative noise-reduction circuits.

155

7

GENERAL ELECTRIC MULTICHANNEL SOUND DECODER

This chapter is devoted to the circuits of a General Electric multichannel sound decoder (also called the stereo/bilingual adapter). The adapter/decoder makes it possible to receive both stereo and SAP broadcasts on a monaural TV set. If the TV set does not have built-in stereo amplifiers and speakers, an external system must be provided. However, the adapter does provide both left-channel and right-channel audio (at a low level).

Unlike the unit described in Chapter 5, the adapter/decoder discussed in this chapter *does not have any operating controls or indicators.* These must be provided on the TV set.

Connections to and from the TV set are made through a phono-type plug (for IF input) and a DIN connector (for audio output and control in/out lines).

7-1. CIRCUIT DESCRIPTIONS

All of the notes described in Sec. 1-7 apply to the following descriptions.

7-1.1 Basic adapter/decoder functions

Note that the adapter circuits perform the same functions as those described for MTS circuits, but the audio paths are quite different. The basic functions performed by the adapter/decoder are as follows (See Fig. 7-1).

The IF signal from the receiver tuner is processed to recover a 4.5-MHz audio signal (which is, hopefully, as void of spurious components as possible).

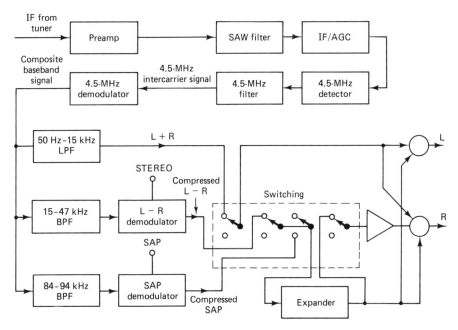

FIGURE 7-1 Basic adapter/decoder functions.

The 4.5-MHz intercarrier signal is demodulated to obtain the composite baseband signal. The baseband contains the L + R mono signal, the L − R stereo signal, the pilot carrier, and the SAP subcarrier. The components of the composite baseband signal are separated by appropriate filtering.

The L − R subcarrier is demodulated, using the pilot carrier to restore the suppressed carrier (to recover the L − R audio signal). The SAP subcarrier is demodulated to recover the SAP signal (if present). D-c levels are developed to turn on indicator lamps that inform the user of stereo and/or SAP broadcasts.

The TV receiver supplies d-c signals to the adapter/decoder to indicate the user's choice of stereo, SAP, or mono operating modes. Depending on user choice, either the L − R stereo signal or the SAP signal is passed through noise-reduction and expander-processing circuits to restore normal audio signal levels. The output of the noise-reduction and processing circuits is matrixed with the mono signal to get the proper outputs for an external stereo amplifier.

7–1.2 Input circuit

Figure 7–2 shows the input circuits of the adapter/decoder in simplified form. A low-level tuner output is coupled from the receiver to the input of the adapter/decoder through a shielded cable, and applied to IF amplifier Q101. The MTS signal (called the MCS signal in General Electric literature) does not pass through the receiver video IF system because of the presence of interfering spurious signals that may fall into the L − R subcarrier frequency range. These signals are generated because of frequency modulation of the AM video signals that occur as the signal passes through a typical video IF amplifier system. Spurious content in the L − R subcarrier could produce a buzz in the recovered stereo audio signal.

The output of Q101 is passed through SAW filter SF101, which has a response to the signal as shown in Fig. 7–3. Note that the video signal is centered on a peak that minimizes undesired FM modulation of the video carrier, and that the audio IF signal is located on a second peak with a pronounced null between the two peaks.

The output of the SAW filter SF101 is applied to IC101. Integrated circuit IC101 contains an IF amplifier, an AGC control circuit, and a product detector system from which the baseband audio signal is obtained. The audio is taken from pin 6 of IC101 and applied to the three filters of the input-filter system through audio-driver/impedance-match transistor Q102. Note that IC101 is tuned by L104 and L105.

7–1.3 L+R audio path

Figure 7–4 shows the L + R audio path in simplified form. Audio from Q102 is applied to Q103 through a 0- to-15-kHz low-pass filter. This filter passes the L + R mono signal but rejects both the L − R and SAP signals.

The mono L + R signal is coupled to the Q103 deemphasis circuit, which reverses the transmitted preemphasis (high-frequency boost) process. The level of the L + R signal is set by R118.

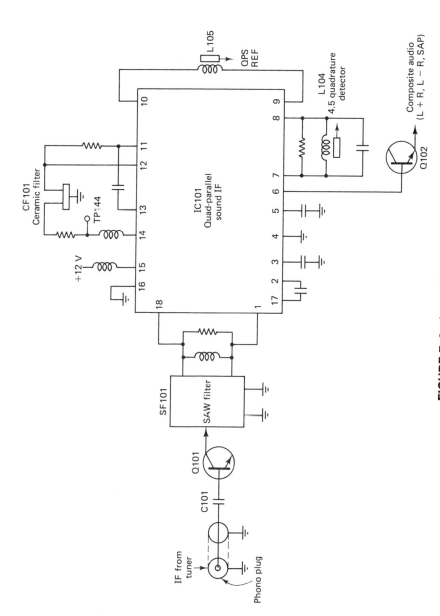

FIGURE 7-2 Input circuit.

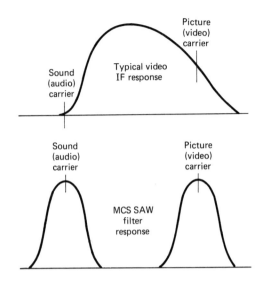

FIGURE 7-3 Comparison of typical video IF response to SAW filter response.

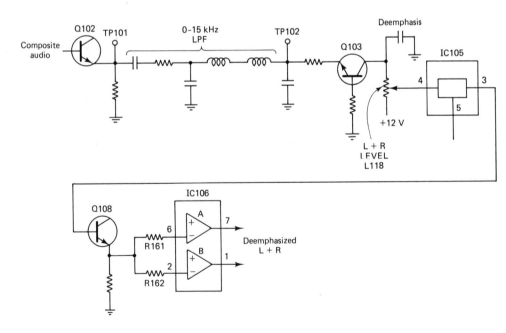

FIGURE 7-4 L+R audio path.

The deemphasized L + R signal is applied to mode switch IC105 at pin 4. The L + R signal exits IC105 at pin 3 (assuming the SAP mode is not selected) and is applied to output matrix IC106 through buffer Q108.

7-1.4 L – R audio path

Figure 7-5 shows the L – R audio path in simplified form. Audio from Q102 is applied to L – R stereo demodulator IC102 through Q104 and a 15- to-47-kHz bandpass filter. This filter passes the L – R signal but rejects both the L + R and SAP signals. The filter is tuned by L109, L110 and L111. The level of the L – R signal applied to IC102 is set by R125.

The L – R signal is demodulated with IC102 by synchronous detection using the 15.734-kHz pilot signal as a frequency reference. (If there is no pilot, there is no detection and no audio from IC102.)

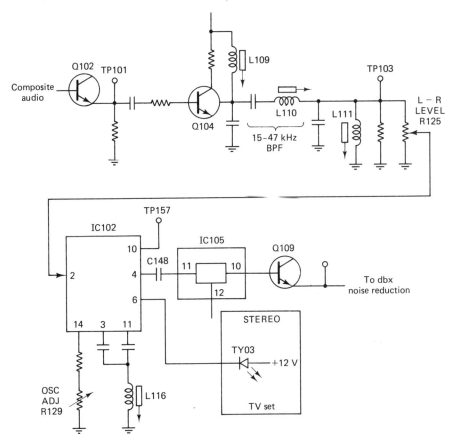

FIGURE 7-5 L – R audio path.

The resultant audio from IC102 is taken from pin 4 and is applied to pin 11 of IC105. Except when SAP is selected by the user, the L−R signal is connected to pin 10 of IC105 and is applied to the noise-reduction and expansion-processing circuits (described in Sec. 7–1.6).

IC102 also contains a trigger circuit which detects the presence of a pilot carrier (stereo transmission). When a stereo pilot is present, pin 6 of IC102 goes low, turning on TV front-panel STEREO lamp TY03. The reference oscillator within IC102 is set by R129.

7–1.5 SAP audio path

Figure 7–6 shows the SAP audio path in simplified form. Audio from Q102 is applied to SAP demodulator IC103 through an 84- to 94-kHz bandpass filter IC BF100. This filter passes the SAP signal but rejects both the L+R and L−R signals. The filter is untuned.

The filtered SAP signal from BF100 is applied to pin 14 of IC103, which demodulates the FM subcarrier and provides the recovered audio output at pin 2. The SAP audio is applied to pin 8 of mode switch IC105 through SAP level control R138. IC103 is tuned by L115.

If the SAP mode is selected by the user, a positive voltage is provided by the TV receiver to turn on Q111. This removes the +12 V from pin 5 of IC105 and disconnects pins 3 and 4, interrupting the L+R audio (Sec. 7–1.3). The +12 V is also removed from pin 12 of IC105, opening pins 10 and 11, to interrupt the L−R audio (Sec. 7–1.4).

With the collector of Q111 low (Q111 turned on), Q107 is turned off, applying +12 V to pin 6 of IC105. This connects pins 8 and 9 of IC105 and applies the SAP audio to expander buffer Q109.

Also note that the adapter/decoder can be placed in a forced-mono or L+R mode by interrupting a positive voltage from the TV receiver. During normal stereo operation, this voltage is applied to pin 12 of IC105 through R152. When forced mono is selected (say, when the stereo signal is too weak to produce good sound), the voltage is removed by the front-panel STEREO switch, pin 12 of IC105 goes low, and pins 10 and 11 are opened to interrupt the L−R audio. Pin 5 of IC105 is kept high by the voltage through R230, so L+R can pass through pins 3 and 4 of IC105.

Approximately 1.5 V of SAP subcarrier signal is present at pin 9 of IC103 (TP106). The subcarrier is detected by a quadrature detector in IC103 (tuned by L115) and applied to Q105 from pin 10 of IC103.

The detected SAP signal is further amplified by Q105 and rectified by Y101/Y102. The resultant d-c voltage turns on Q106, causing the collector of Q106 to go low. This low turns on front-panel SAP lamp TY02. (Note that on some TV sets, the SAP indicator is referred to as the "bilingual" indicator, since one of the intended purposes of SAP is to broadcast in languages other than that of the main broadcast.)

The rectified output from Y101/Y102 is also applied to pin 6 of IC103. If

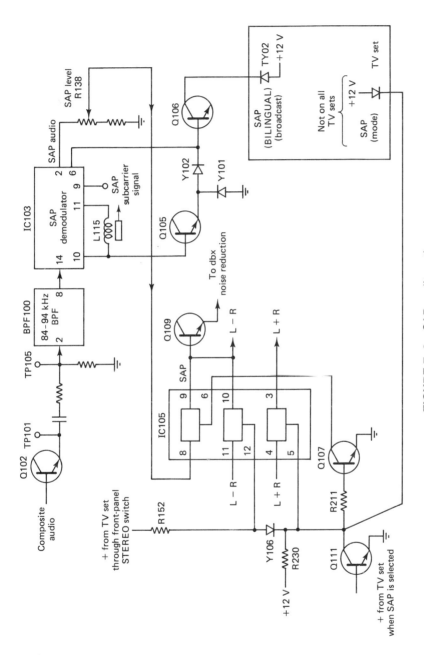

FIGURE 7-6 SAP audio path.

163

there is no SAP carrier, the quadrature detector within IC103 is turned off or muted. The muting circuits eliminate the noise "burst" that would occur if the SAP mode is chosen without a SAP signal being broadcast.

7-1.6   Noise reduction and expansion processing

As in the case of any other MTS circuit, both the stereo and SAP program signals must be processed by the variable deemphasis and expander system to restore normal signal levels and achieve dynamic noise reduction.

Figure 7-7 shows the noise-reduction and expansion-processing circuits in block form. Fig. 7-8 shows the same circuits in simplified form.

Expander IC104 is the heart of the noise-reduction circuits. (Note that expander IC104 is generally referred to as the dbx noise reduction or NR IC in other literature.) No matter what the IC is called, IC104 has two adjustments. *Expander timing control* R191 sets the amount of control output by setting the current through R213 (typically, for 15 mA as measured at TP213/TP214). *Spectral control* R189 ´ sets the desired variable deemphasis by setting L−R/SAP gain, after spectral processing but before wideband processing.

Either L−R or SAP audio is applied to pin 18 of IC104 through buffer Q109 and fixed deemphasis network C156/R178. The selected audio is subjected to variable deemphasis or VD at this point.

The VD process is done primarily by one of the current-controlled amplifiers (CCA) contained in IC104. (Note that the CCA is often referred to as the VCA, or voltage-controlled amplifier, in other literature.) In any event, the gain of the CCA is controlled by an rms detector in IC104. The rms-detector input is a signal at pin 20. This signal is preemphasized by op-amp C of IC106 and buffered by Q112/Q110.

When the input signal to the rms detector contains high frequencies, the highs are boosted by the CCA. When high-frequency audio signals are not present, the amplifier gain is reduced and the high-frequency noise components of signal are minimized.

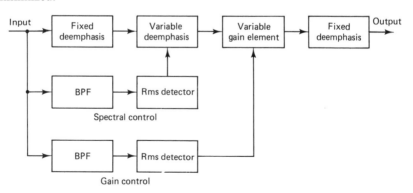

FIGURE 7-7   Basic expansion/noise-reduction process.

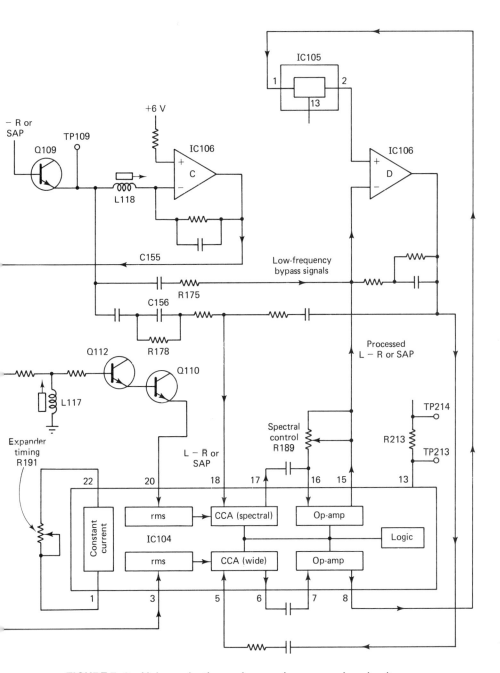

FIGURE 7-8 Noise-reduction and expansion-processing circuits.

Low-frequency components of the signal bypass the variable deemphasis circuit and are applied through C155/R175 to the input of op-amp D within IC106.

The noise-reduced signals that exit the variable deemphasis CCA at pin 17 of IC104 are adjusted for level by spectral control R189 and are returned to an op-amp within IC104 at pin 16.

The amplified output at pin 15 of IC104 is combined with the low-frequency bypass signals, and both are applied to the input of op-amp D within IC106. The output of op-amp D is then applied to pin 5 of IC104, where wideband processing or expansion occurs.

Since the stereo and SAP signals are "compressed" at the transmitter (as a noise-reduction technique), the process is reversed by wideband expansion. The wideband circuits operate in a manner similar to the noise-reduction circuits just described.

The signal at CCA input (pin 5 of IC104) undergoes a *variable degree of amplification,* as determined by the level of the signal applied to the input of the corresponding rms detector (at pin 3 of IC104). High-level input signals are amplified more than low-level signals, resulting in restoration of the dynamic range (hopefully identical to that of the original broadcast audio, before processing).

The audio from the wideband CCA is taken from pin 6 of IC104, returned to an op-amp within IC104 (at pin 7) for further linear amplification, and then taken from pin 8. The output from pin 8 of IC104 is coupled to the inverting (−) inputs of the A and B op-amps within IC106 through the matrix, as shown in Fig. 7–9.

The noninverting (+) inputs of the A and B op-amps receive different signals, depending on which mode is selected. The outputs of op-amps A and B form the left-channel and right-channel audio outputs, respectively.

If the SAP mode is chosen, pins 1 and 2 of IC105 are open, and the input to IC106 is the same as for a mono or L + R signal from Q108 (through the matrix). The recovered (or processed) SAP signal from IC104 is applied equally at the adapter/decoder left-channel and right-channel outputs.

If the stereo mode is chosen, the L − R signal from pin 8 of IC104 is coupled through mode switch IC105, pins 1 and 2, to the noninverting inputs of the op-amps, as well as to the inverting inputs through the matrix.

The mixture of the L + R signals from Q108 to the inverting inputs, and the processed L − R signal to both inputs, results in left-channel audio at pin 7 of IC106 and right-channel audio at pin 1.

7–2. TROUBLESHOOTING APPROACH

All of the notes described in Secs. 1–7 and 1–8 apply to the following procedures. In each of the following trouble symptoms, it is assumed that the adapter/decoder is used with a known-good TV set and a known-good stereo system. That is, it

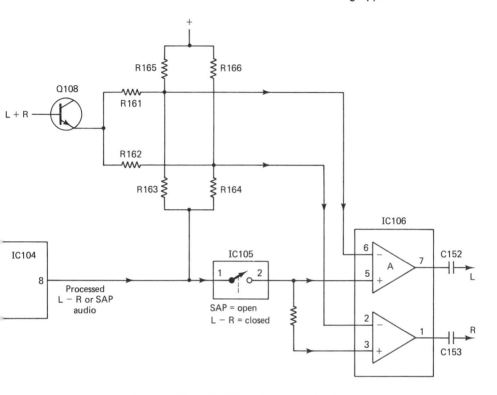

FIGURE 7-9   Matrix and audio output.

is assumed that mono broadcasts can be played through the TV set and audio system, and that problems arise only when the adapter is used.

It is also assumed that an active TV channel has been tuned in, that both stereo and SAP signals are (supposedly) being broadcast (in addition to a mono broadcast), and that the video is good. If you do not have a good picture, the problem is probably in sections ahead of the adapter circuits (such as in the RF or IF circuits).

As a first step, check if any audio is available at the loudspeakers (either at the TV set or external stereo system). If the TV set has STEREO and/or SAP lamps, check that the lamps are on, indicating stereo and/or SAP broadcasts.

7-2.1   No audio available at the loudspeakers

If there is no audio available at the loudspeakers, but the picture is good and the STEREO indicator (if any) on, the problem is probably in the main or L + R audio path, or in the audio amplifiers that follow the main audio path. This can include both the audio amplifiers in the matrix IC (IC106 in Fig. 7-9) and amplifiers in the TV set and/or external stereo. The fact that the STEREO indicator is on makes it likely (but not absolutely certain) that the sections ahead of the adapter are good.

Before you get too far into any of the circuits, make certain that the adapter is receiving proper control signals from the TV set. As an example, to receive mono L + R audio, pin 5 of IC105 (Fig. 7-6) must be high (about 12 V). Pin 5 is normally made high through R230. However, if Q111 is turned on by a SAP control signal from the TV set, pin 5 of IC105 goes low. So if you do not get L + R audio, check that Q111 is not turned on and that pin 5 of IC105 is high.

There are several approaches that can be used at this point. However, there are three logical choices for practical troubleshooting.

1. You can inject an IF signal with L + R at the phono-plug input (at C101, Fig. 7-2).
2. You can inject a composite signal with L + R at pin 6 of IC101 (Fig. 7-2) or the base of Q102 (Fig. 7-4).
3. You can inject audio at the matrix input or output (at IC106, Fig. 7-9).

Let us start with this last approach to clear all of the audio circuits (IC106, TV set, external stereo). Inject audio at pins 1, 2, 3, 5, 6, and 7 of IC106. If there is sound at the loudspeakers with audio injected at pin 1 but not at pins 2 or 3, IC106 is suspect. The same is true if you get sound when audio is injected at pin 7 but not at pins 5 and 6 of IC106.

If there is no sound when audio is injected at pins 1 and 7 of IC106, inject audio on both sides of C152/C153. If there is still no sound, the problem is probably in the audio circuits following the adapter (or possibly bad connections between the adapter and the TV set).

If the audio circuits appear to be good, inject an IF signal modulated with L + R at the phono-plug input (Fig. 7-2). Then check for audio at both loudspeakers. Next, inject a composite signal modulated with L + R at the base of Q102, or at TP101 (Fig. 7-4). Again check for audio at both loudspeakers.

If the modulated signal is heard on both loudspeakers when L + R is injected at the base of Q102 (or TP101) but not when injected at the phono-plug input, the input circuits shown in Fig. 7-2 are suspect. Trace through the signal path using signal injection or signal tracing, whichever is most convenient. The signal path includes IF amplifier Q101, SAW filter SF101, and quad-parallel sound IF IC101. Also, try correcting the problem by adjustment of L104 and L105.'

If the modulated signal is heard on both loudspeakers when L + R is injected at the phono-jack input, it is reasonable to assume that the signal path through the input and L + R circuits (Figs. 7-2 and 7-4) and the audio amplifiers is good. The problem is probably ahead of the adapter/decoder circuits.

If the modulated signal is not heard on the loudspeakers with an L + R signal applied to Q102 and/or TP101, trace through the audio path using signal injection or signal tracing, whichever is most convenient.

The L + R path (up to the matrix) includes Q102, Q103, L + R LEVEL control

R118, IC105, and Q108, shown in Fig. 7–4. Also, try correcting the problem by adjustment of R118.

7–2.2 No stereo operation, mono operation is good

If there is mono audio available at the loudspeakers, and the picture is good but there is no stereo operation, check that the adapter is receiving proper control signals from the TV set.

As an example, for proper stereo operation (both mono L+R and stereo L−R), pins 5 and 12 of IC105 must be high (about 12 V). Pin 5 is normally made high through R230, as discussed in Sec. 7–2.1 (Fig. 7–6). Pin 12 is normally made high through R152. These highs can be removed by either of two conditions, as shown in Figs. 7–4 and 7.5.

If a SAP control signal is applied, Q111 turns on, dropping the collector to ground (actually about 0.2 V). This removes the high from pin 5 of IC105, as well as from pin 12 through Y106. This same condition can occur if Q111 turns on accidentally.

If a forced-mono condition is selected at the TV set, the 12 V is removed from pin 12 of IC105. This interrupts the L−R signal both before and after noise-reduction processing by IC104.

If the TV set has a STEREO indicator, check that the lamp (such as STEREO lamp TY03 in Fig. 7–5) turns on, indicating that stereo is selected and is being broadcast. If the STEREO lamp is off, it is possible (but not absolutely certain) that stereo is not being broadcast! The quickest approach at this point is to inject a composite signal with L−R at TP101.

If STEREO lamp TY03 turns on and you get stereo at the loudspeakers, the problem is probably ahead of the adapter, but could be in the input circuits of Fig. 7–2. (If the input circuits appear to be good for mono operation, as checked in Sec. 7–2.1, it is reasonable to assume that the input circuits are also good for stereo operation.) If STEREO lamp TY03 does not turn on and/or there is no stereo at the loudspeakers, the problem is in the L−R path or the noise-reduction path.

There are several points to consider when tracing through the L−R and noise-reduction paths, with either signal injection or signal tracing.

If STEREO is not on, the problem is probably in Q104, IC102, or the associated circuits shown in Fig. 7–5. Try correcting the problem by adjustment of L109, L110, L111, L116, R125, and R129.

If STEREO is on, it is reasonable to assume that the IC102 VCO is probably functioning properly and reinserting an L−R carrier. However, this should be confirmed by further testing. Trace audio from pin 4 of IC102 to Q109.

If there is audio at pin 4 of IC102 but not at pin 11 of IC105 suspect C148. If there is audio at pin 11 of IC105 but not at pin 10, suspect IC105. It is also possible that IC105 is not receiving proper control signals from the TV set, as discussed in Sec. 7–2.1.

If there is audio at pin 10 of IC105 but no stereo operation, the noise-reduction circuits shown in Fig. 7–8 are suspect. These include Q109, Q110, Q112, IC104, and IC106. Try correcting the problem by adjustment of L117, L118, R189, and R191.

7–2.3 No SAP operation, mono and stereo are good

If there is mono and stereo audio, and the picture is good but there is no SAP operation, check that the adapter is receiving proper control signals from the TV set. As an example, for proper SAP operation, pin 6 of IC105 must be high (about 12 V). As shown in Fig. 7–6, this connects pins 8 and 9 of IC105 and permits SAP audio from IC103 to pass to the noise-reduction circuits at Q109.

Pin 6 of IC105 is normally low, since Q107 is turned on through R211 and R230. When SAP is selected at the TV set, Q111 is turned on, dropping the collector to near zero. This turns Q107 off, causing pin 6 of IC105 to rise to near 12 V. As discussed, when Q111 is turned on, pins 5 and 12 of IC105 go low, removing L + R and L − R.

If the TV set has a SAP indicator, check that the lamp (such as SAP lamp TY02 in Fig 7–6) turns on, indicating that SAP is selected and is being broadcast. If the SAP lamp is off, it is possible (but not absolutely certain) that SAP is not being broadcast.

Note that in some configurations, the collector of Q111 is connected to a SAP mode lamp, as shown in Fig. 7–6. When Q111 is turned on by a SAP control signal, the SAP mode lamp is connected to near ground by Q111, and the SAP mode lamp turns on. However, this is not to be confused with the SAP lamp TY02, which turns on only when SAP is being broadcast. The quickest approach at this point is to inject a composite signal with SAP at TP101 and/or TP105.

If SAP lamp TY02 turns on and you get SAP at the loudspeakers, the problem is probably ahead of the adapter, but could be in the input circuits of Fig. 7–2. If the input circuits appear to be good for mono operation (Sec. 7–2.1) and stereo operation (Sec. 7–2.2), it is reasonable to assume that the input circuits are also good for SAP operation. If SAP lamp TY02 does not turn on and/or there is no ' SAP at the loudspeakers, the problem is in the SAP path or the noise-reduction path.

There are several points to consider when tracing through the SAP and noise-reduction paths, with either signal injection or signal tracing. If SAP is not on, the problem is probably in BF100, IC103, Q105, Q106, Y101, Y102, or the associated circuits shown in Fig. 7–6. Try correcting the problem by adjustment of L115 and R138.

If SAP is on, it is reasonable to assume that IC103 is probably functioning properly. However, this should be confirmed by further testing. Trace audio from pin 2 of IC103 to Q109.

If there is audio at pin 9 of IC105 but no SAP operation, the noise-reduction circuits shown in Fig. 7–8 are suspect. These include Q109, Q110, Q112, IC104, and IC106. Try correcting the problem by adjustment of L117, L118, R189, and R191.

If the noise-reduction circuits appear to be good for stereo operation (Sec. 7–2.2), it is reasonable to assume that the noise-reduction circuits are also good for SAP operation.

INDEX